Philosop
Part II:

Book Series
Philosophy for Heroes

PHILOSOPHY FOR HEROES

PART II: CONTINUUM

Published by Clemens Lode Verlag e.K., Düsseldorf

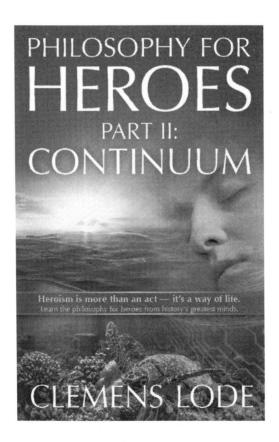

2018, First Edition

ISBN 978-3-945586-22-8

Edited by: *Conna Craig*
Cover design: *Jessica Keatting Graphic Design*
Image sources: *shutterstock, istockphoto*
Icons made by http://www.freepik.com from
http://www.flaticon.com is licensed by CC 3.0 BY
(http://creativecommons.org/licenses/by/3.0/)

Printed on acid-free, unbleached paper.

Subscribe to our newsletter. Simply write to newsletter@lode.de or
visit our website https://www.lode.de.

PHILOSOPHY
POPULAR SCIENCE
PSYCHOLOGY

Dedication

As we begin the next stage of examining what it really means to be a hero, let us reflect on my favorite quote from Lawrence Krauss:

 That, finally, is the most accurate picture I can paint of reality as we now understand it. It is based on the work of tens of thousands of dedicated minds over the past century, building some of the most complex machines ever devised and developing some of the most beautiful and also the most complex ideas with which humanity has ever had to grapple. It is a picture whose creation emphasizes the best about what it is to be human—our ability to imagine the vast possibilities of existence and the adventurousness to bravely explore them —without passing the buck to a vague creative force or to a creator who is, by definition, forever unfathomable. We owe it to ourselves to draw wisdom from this experience. To do otherwise would do a disservice to all the brilliant and brave individuals who helped us reach our current state of knowledge.

—Lawrence Krauss, *A Universe From Nothing*

Introduction

Εύρηκα!—*Eureka!*—Greek, roughly meaning "I have found (it)!"—is a reference to the ancient Greek scholar Archimedes. He discovered he could precisely determine the volume of any object by submerging it in water and measuring the amount the water rose. He reportedly shouted "Eureka! Eureka!" and leapt out of his bathtub and ran through the streets of Syracuse naked to share his discovery of this previously unsolved problem. The expression is a synonym for a sudden realization, a previously undiscovered connection between two concepts leading to a new insight. It is also the state motto of California. While as a motto it references the discovery of gold in 1848, it can also be seen as a symbol of scientific discovery.

THIS BOOK SHOWS some of the Εύρηκα!—*Eureka!*—moments of science. It will point out the moments when a new connection between scientific concepts led to a massive simplification of the sciences. From planetary movements, a deterministic interpretation of quantum theory, evolution, the origin of life, up to human creativity, these *Eureka!* moments provide you the courage to see the world as something you are able to understand. The book series *Philosophy for Heroes* offers the intellectual and moral know-how necessary to withstand a chaotic world in a responsible manner, to encourage you to share the knowledge gained here, and to become a shining example for others.

Contents

Publisher's Note

 Since the launch of the first book of this series, *Philosophy for Heroes: Knowledge*, a lot has happened. I have optimized my work processes, managed to secure financing by management consulting projects about leadership and organization, rented a tiny office, and released two books on project management.

Now, I am ready for the next stage.

THE BIGGEST CHALLENGE now is to decide whether to hire outside help for some parts of the books, and to determine which parts to research and write myself. The goal is to not spend too much time on the details while focusing more on what matters to you, the reader. I wrote earlier that I see myself as a dreamer; now, a new role is emerging: a leader.

My dream remains the same: a better world, one in which a philosophy of the leader as a hero is embraced and celebrated. My goal is to find more people interested in my dream and to build a community in which they can bring their skills to build something greater together.

But to make this dream a reality, a lot of detail work has to be done by my editor and me, and this especially involves you, the reader. She and I are interested in your comments, thoughts, additions, and in any positive examples of you using the guidelines in this book. To send general feedback, mention the book title in the subject of your message and simply send it to feedback@lode.de. You can also contact us at https://www.lode.de/contact at any time if you are having a problem with any aspect of the book, and I will do my best to address it. Also, I cordially invite you to join my network at https://www.lode.de.

Although I have taken every care to ensure the accuracy of the content, mistakes do happen. If you find a mistake in one of my books, I would be grateful if you would report this to me. By doing so, you can help me to improve subsequent versions and maybe save future readers from frustration. If you find any errata, please report them by visiting https://www.lode.de/errata, selecting your book, and entering the details. Once your errata are verified, your submission will be accepted and the errata will be uploaded to my website, or added to any list of existing errata, under the Errata section of that title. You will, of course, be credited if you wish.

If you are interested in the process of how this book was produced, take a look at chapter 4.7. There, I explain how I applied the lessons I learned from writing the first book.

Thanks to *you, the reader*, the series can continue. Thank you for keeping up the tradition of reading books and supporting the project by your interest in this topic. You and your fellow readers have created a market for this book. I hope that I can meet your expectations and I am looking forward to feedback from your side, no matter whether it is positive or negative. This is the only way the project can grow.

- What did you genuinely like about the book?
- What can I improve?
- What do you find unique about the book?
- What subjects would you like to read more about?
- Tell your story! How did you find the book?

Do you have a different view? Let us know what you thought about the book—whether you liked it or not—via contact@lode.de or directly via my network https://www.lode.de/contact. This will help us to develop future titles.

Best regards,

Clemens Lode
Düsseldorf, Germany, February 1st, 2018

Preface

" When I was alive, I believed—as you do—that time
was at least as real and solid as myself, and probably more so. I said "one o'clock" as though I
could see it, and "Monday" as though I could find
it on the map; and I let myself be hurried along
from minute to minute, day to day, year to year,
as though I were actually moving from one place
to another. Like everyone else, I lived in a house
bricked up with seconds and minutes, weekends
and New Year's Days, and I never went outside until I died, because there was no other door. Now I
know that I could have walked through the walls.
[...] You can strike your own time, and start the
count anywhere. When you understand that—
then any time at all will be the right time for you.

—Peter S. Beagle, *The Last Unicorn*

In the first book, *Philosophy for Heroes: Knowledge*, we learned that being a hero means to stand up against false heroes, but also against false teachings. We can gain knowledge about the world only by going back and forth. New ideas lead us to re-examine old ones, which again can lead to new ideas. This is especially true when discussing topics like cognition and everything that is connected to it. In the first book, we started out with an intuitive understanding of the world consisting of entities which we can objectively perceive. In this book, we re-examine this idea on the physical level, as well as prepare the way to test it on the level of our own consciousness, which we will discuss in depth in the following book, *Philosophy for Heroes: Act*.

What you have learned in the first book—the foundations of philosophy, language, and mathematics—will help you to better understand this second book. If you are starting with *Philosophy for Heroes: Continuum*, you will find relevant concepts from the first book explained herein.

My goal with the *Philosophy for Heroes* series is to give you a comprehensive overview from the very fundamental ideas of ontology ("what is?") and epistemology ("how do we know?") up to the fields of psychology, ethics, and leadership. This knowledge allows you to better reflect who you are and to be original—a leader. The first book dealt with the topic mainly from logic and our conceptual understanding of the world. The third book will deal with the topic mainly from examining our consciousness and how our mind works. This book bridges both insofar as it deals with science, how reality is more like a continuum than a distinct set of entities, the gradual emergence of life from the non-living universe, and the evolution of complex systems up to our own intelligence and creativity.

Becoming a hero requires knowing where you come from. It requires seeing your own place in the universe and how you connect to it and all the living things around you.

The topics discussed here are challenging. The entire book series can be viewed as a system of interwoven ideas, some of which can become more clear as we move through topics that may at first seem disconnected. Thus, I ask you for patience—if some expression or statement meets your disapproval, consider it within a larger scope of ideas. If we were to show many of today's dominating worldviews on a canvas, some of the topics discussed would likely appear to be out of line. The purpose of *Philosophy for Heroes*, however, is not to repair prevailing worldviews, but to start afresh on a completely new canvas, and comprehend the world from a different perspective.

I will attempt to present this new perspective by building bridges to other philosophies and to science. While I have cited works of others frequently, this book series will present my own point of view. Especially with remarks I have made without references, these are my own ruminations and may be taken merely as starting points for the reader to think about further. The compact nature of *Philosophy for Heroes*, as it touches on a great number of topics, requires this approach.

That being said, just like movie trailers or the first level of a computer game, prefaces are never written at the beginning but at the end, when the artist or writer has acquired the necessary skills and knowledge to complete the project. A creator emerges from his or her creation transformed. As such, this book series is also like a diary of my own studies and self-development.

My hope is that you will take with you from this book series one or two interesting thoughts, and develop them further or let them inspire you. Personally, I would like to place this book in the hands of a younger version of myself, someone who finds himself at the beginning of his journey of scientific discovery, having to first wade through tons of misinformation before getting to the "good stuff." Even if I reach only a small handful of people who will take to heart a few of these core ideas and set out to achieve something great in life themselves—whatever that might be—I will be able to enjoy the rewards of this book. I made the book available to the public just as I would plant a handful of seeds in the earth, hoping that they grow and bloom.

You, the reader, are holding the open book in your hands and now have an idea of what to expect. Let us continue our journey to become a shining example for others, thus getting closer to our ideal world. The path to our goal is the *Philosophy for Heroes*.

Clemens Lode
Düsseldorf, Germany, February 1st, 2018

You can find chapters 1 and 2 in the previous book,

PHILOSOPHY FOR HEROES

PART I: KNOWLEDGE

Published by Clemens Lode Verlag e.K., Düsseldorf, Lode,
Philosophy for Heroes: Knowledge

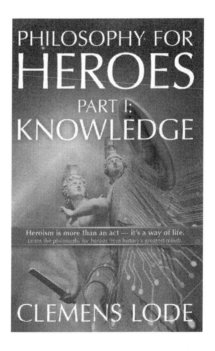

Chapter 3

Physics

“ The ideal subject of totalitarian rule is not the convinced Nazi or the convinced Communist, but people for whom the distinction between fact and fiction (i.e., the reality of experience) and the distinction between true and false (i.e., the standards of thought) no longer exist.

—Hannah Arendt, *The Origins of Totalitarianism*

T HE WORD "science" is Latin (*scienzia*) for knowledge. But it means so much more than what we covered in the first book. There, we discussed our cognition and how we can conceptualize the world. Among the central concepts we have discussed are:

> **ENTITY** · An *entity* is a "thing" with properties (an identity). For example, a plant produces oxygen, a stone has a hard surface, etc.).

> **IDENTITY** · An *identity* is the sum total of all properties of an entity (e.g., weight: 160 pounds, length: 6 feet, has a consciousness, etc.).

> **PROPERTY** · A *property* refers to the manner in which an entity (or a process) affects other entities (or other processes) in a certain situation (e.g., mass, position, length, name, velocity, etc.).

> **CONFIGURATION OF A PROPERTY** · The *configuration of a property* relates to the intensity of a certain property of an entity.

> **EFFECT** · An *effect* is the change caused to the configuration of the properties of an entity (e.g., the heating of water changes its temperature).

> **PROCESS** · A *process* describes the mechanism of a cause working to an effect (e.g., if you put an ice cube into a glass of water, the cooling of the water is the process).

Here, we actively create experiments to discover more about the world. This approach is the path of a hero: not to be a passive bystander, but instead to interact with his or her environment. This interactive approach also poses new challenges where we no longer can simply look at entities (and ourselves) being strictly separated from each other. Instead, we have to look at a *continuum* of entities. Instead of looking at reality as a series of snapshots or discrete entities, we are looking at infinitely folded recursive processes. Many processes in nature cannot be understood by purely entity-based thinking and language. We say that "it is raining" instead of talking about the positions of individual raindrops. In this book, we will talk about a number of such processes, from quantum theory to evolution. Nature is not concerned about our predilection to dividing it into small parts in order to understand it, nature is a *continuum.*

> The question now is, how does it really work? What machinery is actually producing this thing? Nobody knows any machinery. Nobody can give you a deeper explanation of this phenomenon than I have given; that is, a description of it. They can give you a wider explanation, in the sense that they can do more examples to show how it is impossible to tell which hole the electron goes through and not at the same time destroy the interference pattern. They can give a wider class of experiments than just the two slit interference experiment. But that is just repeating the same thing to drive it in. It is not any deeper; it is only wider. The mathematics can be made more precise; you can mention that they are more complex numbers instead of real numbers, and a couple of other minor points which have nothing to do with the main idea. But the deep mystery is what I have described, and no one can go any deeper today.

—Richard Feynman, *The Character of Physical Law*

Question

Why should one state follow from another at all? Why is everything not "frozen?" Why do entities act according to their properties all the time? Could they not act in one year like this, and in another year like that?

To answer these questions, let us take a step back. In *Philosophy for Heroes: Knowledge*, our approach to knowledge required the universe to be deterministic. A deterministic view of the universe means that one (and only one) state follows from a previous state: every event has a cause, and every cause is an event. Before we discuss the findings and interpretations from physics, let us re-examine the concept of determinism on a purely philosophical level. Causality means that effects have causes. That means that the properties of an entity are realized. And the only way "nothing" would follow an entity is if the entity ceases to exist.

> **CAUSALITY** · *Causality* refers to the effect of one or several entities on another entity in a certain situation (e.g., an accident is no random occurrence, there are one or several causes which led to the accident, such as lack of sleep, a technical defect, poor visibility, etc.).

Example

One property of an entity could be that its other properties change over time. This is something we can observe. We might need a lot of time to observe seemingly random changes and find their underlying pattern, but we would find them eventually. A current example for this is *Tabby's Star*, a star 1,480 light years from Earth. With the help of telescopes, a group of amateur scientists have noticed a strange, seemingly random pattern of changes in brightness. At this point (2018), it is unclear what is causing it. For example, orbiting planets would show a very regular pattern of changes in brightness. All options are open, from planets with ring systems like our Saturn, huge artificial (alien) structures, irregular dust clouds, or large comet showers. But we know there must be a cause and it is not the sun because the sun does not violate its own properties.

Now, let us look at the opposite scenario. If our universe was not deterministic, what would that mean? In such an indeterministic world, entities would not act according to their identity. The effects we could observe would no longer help us to identify their causes and thus the underlying properties of an entity. In addition, in an indeterministic world, our own perceptual faculty and cognitive system would be based on indeterminism. This could still "work" if it is merely small particles jiggering randomly. In large numbers, the randomness of individual particles would not matter.

Imagine a large choir, singing a song. Even if a few people sing the wrong words or sing in the wrong tune, you can still hear the actual song. It could also be compared to radiation: our body can handle some radiation without problems. But eventually, the damage due to random destruction of individual cells adds up. Likewise, this applies to a higher, logical level as well. If we cannot be sure that the words we say or understand from another person are really the words we wanted to say or that the other person said, a conversation is impossible.

Idea

If everything were random, something like the human brain could not exist. Such complex structures can emerge only when there is some sort of pattern in the laws of nature. So, the question is not whether the universe is deterministic or not, but how "indeterministic" (if at all) it is. Evidence shows that the degree of "indeterminism" or jiggering in our universe is relatively small. That we exist is proof that stable systems like our bodies can form despite the jiggering at the particle level. The question remains whether we live in an indeterministic world that allows a certain level of order at the macroscopic level (biology), or if we live in a fully deterministic world.

3.1 An Introduction to Science

> It is impossible for someone to dispel his fears about the most important matters if he doesn't know the nature of the universe but still gives some credence to myths. So without the study of nature there is no enjoyment of pure pleasure.

—Epicurus, *Principal Doctrine 12*

Question

Why are ontology and epistemology simultaneous?

As Epicurus put it wisely, for someone to enjoy life, it is necessary to also study what goes far beyond the knowledge required for daily life. It is not enough to be able to refute arguments or dismiss questions about the world. If we do not explore the unknown, that unknown will cause us to be superstitious or will gnaw on our self-confidence. It is not enough to be able to tell what something is *not*. Simply pointing out that something contradicts the existing body of knowledge does not explain our own role in the world; this can leave us feeling unsettled.

Luckily, at this point, we can build upon a foundation from *Philosophy for Heroes: Knowledge*. Science is but a *branch* of philosophy and we can use the results from our studies of philosophy as a basis for our scientific inquiries. We can go from the bottom (philosophy) up (to science), but we *cannot* go from the top to the bottom. If a scientific experiment somehow refutes the very philosophy we are using to conduct the experiment, we must have made a mistake somewhere along the way.

On the other hand, ontology ("what is") and epistemology ("how do we know") are intertwined. New insights into how our own cognition works and how we interact with the world *can* lead to changes in our philosophy. If a new discovery leads to changes of philosophy, experiments that depended on our previous (false) assumptions have to be repeated until both philosophy and science are again in harmony.

Example

In Ancient Greece, people believed rays emitted by our eyes caused us to see. It took until the 10th century, when the astronomer and physicist Ibn al-Haitham invented the first pin-hole camera, to get an idea of how our eyes (and ultimately our visual perception) actually work. A new insight of "what is" led to a change in "how do we know."

Idea

Ontology and epistemology are simultaneous—what exists and how we know it form a foundation of philosophy.

This is the philosophical background one must keep in mind when looking at *interpretations* of empirical data. When such interpretations result in (apparent) contradictions, then we have to check our basic philosophic premises. We do not need to spend time trying to force contradictory interpretations into our philosophical or scientific systems.

When it comes to contradictory *data*, though, that contradicts results from previous scientific experiments, this is something else altogether. Here, we need the *scientific method* in order to sort out falsehoods from truths. That is the task of science and also the main difference between science and philosophy: in science, we can rely and build upon on repeatable experiments; in philosophy, we either get the whole tree of knowledge right or wrong, and then try to sharpen our whole view of the world iteratively. We cannot divide philosophy into independent parts and run experiments on them.

> **SCIENCE** · *Science* is the formalized process of gaining new knowledge from observation, deducting new knowledge from existing knowledge, and checking existing knowledge for contradictions.

3.1.1 Before the Scientific Method

Question

What situation led to the first documented application of methodical scientific research?

Science has played a role in the lives of humans since they began building tools more than three million years ago. Techniques were learned, presented, taught, tested, and further developed. Using only trial and error, the progress was limited to directly applicable and testable knowledge. For example, a production process that resulted in a spear that flew farther and more accurately than previously used spears was copied while others were discarded.

While this type of "pre-science" did not follow a formalized progress like modern science with its theories, it followed the principle of making an assumption of reality (thinking that a certain tool or weapon could solve a problem), and then testing that assumption in the form of an experiment, with others trying to copy the result on their own, creating similar tools.

Did you know?

Compare this approach with rituals before a hunt or harvest. Without a formalized process of conceptualization, testing, and analysis, selective perception caused people to amass "knowledge" about connections in the world that was objectively false. Nobody tried skipping rituals for a few years to test whether the hunt or harvest was affected. Instead, the issue was addressed in an intuitively human way: attributing a consciousness to the world, and then seeing a sacrifice to the world as a form of trade.

$\longrightarrow$ Read more in *Philosophy for Heroes: Epos*

Tool production aside, the real birth of methodical science can be found in medicine. While the use of plants, dental work, wound treatment, etc. probably have a long history, the techniques were taught only verbally through stories. For a written record, we have to go forward to 1550 BC: the Egyptian *Edwin Smith Papyrus*[1] is the first known medical text dealing with wound treatment. At that time, large-scale war had become a part of civilization. Armies with mass-produced weaponry became commonplace, which also meant that on the battlefield, there were thousands of similar injuries. This paved the way for systematic medical research as—just like with the manufacturing of a spear—different treatments could be tested, copied, and refined. Records made it possible to connect the injury with the treatment and the outcome.

Idea

Large-scale war with mass-produced weaponry led to similar injuries on many battlefields. This allowed systematic medical research with different treatments.

---

[1]cf. Aboelsoud, *Herbal medicine in ancient Egypt*.

3.1.2 Placebo Effect

Question

How can we determine if a medical treatment was effective, not just the body healing itself?

This research did not help with individuals where it was difficult to determine the connection between treatment and success. One had to rely on hearsay to select a treatment. If the patient survived, the doctor tried to apply a similar treatment for the next patient, not knowing whether the previous patient healed on his or her own or whether the treatment was actually effective.

That, again, lead to pseudoscience. If the "successful" treatment included praying to the gods, dancing around the patient, pouring "holy water" on the patient, and giving the patient a special herb to eat, the ancient doctors tried to repeat exactly that ritual, unsure which action caused the healing. The doctors also had no formal method of describing their treatments or noting down the statistics of success and failure. And, ultimately, where was the drawback of praying to the gods, using lucky charms, or reciting a "magic spell"?

> **Reversal of the Burden of Proof** · Using the argument of the *reversal of the burden of proof*, you try to evade the necessity to give proof for your own arguments and instead present the opposite of your argument, and ask the other person for proof. The basic (and wrong) premise of the reversal of the burden of proof argument is that anything that cannot be disproven must be true. This is a fallacy because you often cannot prove a negative without being omniscient.

Praying, lucky charms, and magic spells do not cause harm and do not influence the outcome, so it was hard to eliminate them from the whole ritual which might have contained one element that helped the healing process. Without a rigorous scientific process to determine a causal relationship, it was hard to differentiate the ingredients of a treatment that actually helped and practices that were added due to ritual.

In that regard, one of the main challenges those early doctors faced was that the body has the ability to heal itself. This lead to "false positives," meaning they misattributed healing to their own actions instead of the body. Ultimately, the "placebo effect" might have played the biggest role in early medical practice. Yes, praying, singing, and dancing around the patient might not have had a direct effect on the illness of the patient, but if the patient just *believed* he would get better, it actually helped.

Modern studies show clearly the biochemical effects of the placebo effect in the brain of a patient.[2] Studies have shown that just fifteen minutes of soothing music significantly lulls a patient so that pain medication can be reduced.[3] Either way, we have to ultimately say that the ancient methods were not *scientific*. There was a lack of understanding of *why* something worked out well. Without that knowledge, the techniques could be applied only to similar cases. Addressing new forms of illnesses required starting from scratch.

Remembering what we have said about the conceptual tree of knowledge, we can clearly see that little to no conceptualization, and thus understanding, about medicine occurred during that time. A method based on trial and error can quickly produce results, but without understanding, they are difficult to apply to future situations. Still, we must not simply dismiss those early practices. In-

---

[2] cf. Colloca and Benedetti, *Placebos and painkillers: is mind as real as matter?*
[3] cf. Mithen, *The Singing Neanderthals—the Origins of Music, Language, Mind, and Body*, p. 96.

stead, we can learn from them and re-examine our own medical research: not that we should also start dancing, but that we need to make sure that the patient is convinced that the treatment will work.

> **Idea**
>
> A method based on trial and error is difficult to apply to future situations if there is no comprehension of what actually worked. In medicine, because of the placebo effect, even without knowing what worked, a positive belief alone can produce results. Likewise, without such a belief, even a comprehension of what worked can lead to a treatment failure.

3.1.3 The Scientific Method

Now I am going to discuss how we would look for a new law. In general we look for a new law by the following process. First we guess it. Then we compute the consequences of the guess to see what would be implied if this law that we guessed is right. Then we compare the result of the computation to nature, with experiment or experience, compare it directly with observation, to see if it works. If it disagrees with experiment it is wrong. In that simple statement is the key to science. It does not make any difference how beautiful your guess is. It does not make any difference how smart you are, who made the guess, or what his name is—if it disagrees with experiment it is wrong. That is all there is to it.

—Richard Feynman, *Messenger Lectures, The Character of Physical Law*

Question

What major problems stood in the way of early researchers before the scientific method?

The term "scientific method" was primarily influenced by the philosopher Karl Popper, who built the method upon the works of previous philosophers, back to Aristotle. But the Greek tradition was certainly not the singular reason that the Scientific Revolution started in Europe and not in Asia. From Asia we inherited inventions like soap, the crank-shaft, quilting, the pointed arch, surgical instruments, windmills, the fountain pen, the numerals, checking accounts, gun powder, and last but not least the idea of having beautiful gardens.

But after the Mongol invasion, large parts of Asia fell into a period of isolationism, while in Europe, religious instability allowed new ideas to flourish. Ironically, one of the reasons for this instability was caused by the Black Death, the spread of which could have been reduced by better sanitation, the availability of soap, and scientific literacy. In a way you could argue that religion no longer was able to provide a safe haven for the people so they turned to studying science. The scientific method took the place previously held by religious authority: theories no longer required an authority, anyone could recreate the experiments and judge the results for themselves.

To understand the scientific method, you need to understand that science itself is a game. If you want to take part in it, you have to follow its rules: the *scientific method*. You can, of course, make discoveries using different rules. Those discoveries could very well be applicable and even successful, but they would not be *scientific*. And that is OK. Just because the scientific method has such a good reputation does not mean you have to do everything in your power to take part in the game of science. That said, without applying the scientific method, you cannot claim that your results are "scientific."

To understand the scientific method, it is best to not just learn its parts, but instead to find out why each part makes sense and what obstacle it solves. Let us look at four major problems that stood in the way of early researchers before the scientific method:

1. People attributed phenomena to a world of ghosts, gods, or dreams.

2. People either rushed to conclusions, or research projects ran forever because people became enamored with their findings or hesitated to admit a failure.

3. People were biased about their findings.

4. People did not share or document their work.

Idea

Before the scientific method, research was limited by bias, a reliance on supernatural explanations and premature conclusions, the absence of a scientific community, and the lack of willingness to admit failures.

Question

What was new about the scientific method?

The first step of the scientific method is to **observe nature** and be curious instead of just accepting the status quo. Asking "why?" about phenomena that are widely accepted is where scientific progress begins. And do not just ask, "Why?" Ask, "Why why why why why...?" until you have found the root causes of a problem or phenomena.[4]

The second step of the scientific method is to gather data and **guess** what the connection between the observed phenomena is. But just stating your conclusions about how the world works is not enough, you need to formulate a hypothesis. A hypothesis is more than a statement based on your observations, it needs to be posed in an if/then form to clearly define the scope and must not have a predetermined outcome. The conclusion must not be already set before the actual investigation begins, as this would harm the objectivity of the research. For example, instead of starting with "all swans are white"—which would raise the question to which swans you are referring—you say "in the summer, all swans in the lake downtown are white." People can then go to the lake, observe what type of swans land there for one season and either support or refute your hypothesis. If all you have is a vague idea about how the world works, you would simply just add new evidence as "special cases" but never really go back to the drawing board to fundamentally examine your views.

Third, too easily, one can become enamored with one's findings. With all the time invested, it might be hard to stop when the initial research does not show the desired results. If you do not clearly define when a research project needs to be curtailed, it might run forever. For this, an additional **null hypothesis**, a definition of when the experiment to test the hypothesis has failed, needs to be defined. It states that what you have observed before were simply random occurrences.

---

[4]Ohno, *Ask why five times about every matter.*

The opposite, the previously mentioned hypothesis, is called the **alternative hypothesis**. It is the connection between different phenomena that you expect to be true based on earlier observations. It is a possible explanation for the observed phenomena. For example, a null hypothesis could be "The water quality did not change over the past 10 years," while the corresponding alternative hypothesis could be "The water quality improved over the past 10 years." To prevent your ego from ultimately being detrimental to your research, it is important to make a very specific prediction to define what exactly you think is true about the alternative hypothesis, as well as a very specific and detailed experimental design to help you to either support or refute the alternate hypothesis. If the prediction with the specific experimental design does not come true, the hypothesis is *wrong*, no matter how beautiful the idea was and no matter what title is held by the person who provided the hypothesis.

Fourth, the experiment has to be run and the **data collected** and **analyzed**. The scientific method here directs scientists to other fields, like logic or statistics, to remove the "intuitive" evaluation and replace it with an objective one. There are hundreds of logical and statistical fallacies that each address an intuitive interpretation. Sometimes, there is simply just a correlation between two phenomena, but no real causation.

Example

Studies show that people who drink wine live longer. But this is only a *correlation*. The real cause might simply because more social and more healthy people tend to drink wine. They might live even longer *without* the wine.

As the fifth and final step, a **conclusion** must be given. You have to be so objective to look directly at the result and accept the null hypothesis if the experiment results do not fit the prediction, instead of trying to manipulate the experiment to show what you had originally anticipated. This objectivity can best be achieved by **documenting** and **sharing** your work with peers who will then try to recreate your experiment given the described conditions.

> The process known as the Scientific Method outlines a series of steps for answering questions, but few scientists adhere rigidly to this prescription. Science is a less structured process than most people realize. Like other intellectual activities, the best science is a process of minds that are creative, intuitive, imaginative, and social. Perhaps science is distinguished by its conviction that natural phenomena, including the processes of life, have natural causes—and by its obsession with evidence. Scientists are generally skeptics.

—Neil A. Campbell, *Biology*

> Science is a collaborative enterprise spanning the generations. When it allows us to see the far side of some new horizon, we remember those who prepared the way [...]

—Carl Sagan, *Cosmos: Blues for a Red Planet*

With these ideas in mind, it is better understood how superstitions of the Middle Ages and ancient times were so popular. People not only had all the difficulties we face today discovering the truth, but they also had no scientific method on which to rely.

Did you know?

Interestingly, when examining our process of cognition, we discover that it is based on principles that are very similar to the scientific method. We make observations, discover unknown information, integrate that into existing knowledge, and try to make connections (form a concept). In order to be sure that our thought process was correct, we reflect upon it, maybe even multiple times, using multiple experiences or the help of others.

$\longrightarrow$ Read more in *Philosophy for Heroes: Knowledge*

Did you know?

Other than external factors, an additional reason the Scientific Revolution started in the West might have been the different philosophical approach to the universe. While Eastern philosophy is focused on a holistic view of the world, in the Western world, the universe was seen as something constructed and researchers saw as their task to uncover "God's plan"—like a machine, if you only knew its parts, you would know the whole thing. In order to manage the complexity of nature, they divided the universe into entities and studied them separately. The scientific method even demands that the observer is separate from the universe, so as not to disturb the experiment. It was only with the dawn of quantum theory that science re-integrated a more holistic view of the universe.

$\longrightarrow$ Read more in *Philosophy for Heroes: Knowledge*

> **Idea**
>
> What was new about the scientific method was that this process was formalized and enhanced by peer reviews from a scientific community in order to reduce bias. Not only can your current peers test your conclusions, but also all future scientists can take your paper from the archives and retest the assumptions. Proper documentation also includes proper citations. A study whose results you have used might turn out to be erroneous. If you have properly cited that study, other scientists can try to correct your work more easily. This openness—having the courage to admit mistakes—is the real driver of scientific progress. Instead of building a knowledge hierarchy of things that are possible to prove, science constructs a knowledge hierarchy where each part openly states that it might be wrong if certain conditions are met. This way, any scientific theory can be proven false by experiment, so any theory building upon other theories always carries with it a whole tree of falsifiable experiments as a prerequisite. In that regard, the start of rigorous application of the scientific method was as significant as the invention of writing. While researchers published books before, there was no process in place to organize this knowledge or access it in a structured manner, like tracing the references back to the original study. With the scientific method, scientists were able to efficiently organize knowledge, and to *trust* and build upon the results of other scientists for the first time in history.

3.2 Occam's Razor

> **Question**
>
> Given that for a phenomenon, explanations of arbitrary complexity can be made, which view of the world should we choose?

OCCAM'S RAZOR PRINCIPLE · According to the *Occam's Razor principle*, among competing hypotheses, the one with the fewest assumptions should be selected. It is attributed to the English Franciscan friar, scholastic philosopher, and theologian William of Ockham (1287 - 1347).

Unlike in pure logic or mathematics, theories in natural sciences are much more difficult to test. You have to find specific experiments which have no or just a few unknown (or uncontrolled) variables in order for them to be repeated. For example, this is a difficult topic for any study involving people because you cannot simply create a "clone" of someone and repeat an experiment. Instead, you need to track gender, lifestyle, age, income, genetics, etc. to at least find some correlation and eliminate unwanted influences. And sometimes experiments are outright impossible because of principle,[5] technological, or monetary limits; think for example about weather or climate scientists who cannot run their experiments on a "second Earth" but have to rely on models.

COMPLEX ARGUMENT FALLACY · It is a *Complex Argument Fallacy* if you use a more complex argument (that leads to the same conclusions as the more simple one) to argue against an existing argument. The new argument requires more assumptions to be true. An extension of this argument is to simply declare a problem or situation as "complex" while dismissing simpler solutions.

---

[5]We will see this later in chapter 3.4 with Heisenberg's Uncertainty principle

> For example, a child could declare that her situation is "complex," complaining she is running out of pocket money, while the reasons are very clear (her expenses). This way, the chance to even discuss a subject is negated instead of addressing actual issues. This fallacy is also often used in connection with an argument from authority, implying that only "experts" are allowed to have an opinion on the subject. Of course, sometimes, there are no simple solutions and a problem is complex, requiring experts to discuss. That is why the complex argument fallacy is a reminder to reflect before either calling a situation too complex, or trying to apply basic solutions that are too simple.

This difficulty of following the scientific method automatically leads to different theories for phenomena. This is unlike what we see in mathematics, or chemistry: there are not many alternative interpretations of mathematics or chemistry, but there are many competing interpretations of psychology, philosophy, biology, and physics (in terms of interpretation of quantum mechanics).

Thus, if you are faced with different explanations for the same phenomena, you should decide in favor of the one that adds the smallest number of hypotheses (assumptions about how the world is) or variables. The point is that for any phenomena and theory, you can always find a more complex theory that explains the phenomena but adds additional, very possibly superfluous or simply incorrect elements. Without some kind of guide or limit, you will never make progress. In addition, any explanations that cannot be falsified—we remember Carl Sagan's invisible, weightless dragon from the first book[6]—should be discarded right away: someone can always try to evade rational inquiry in the form of falsifiable experiments by coming up with more and more fantastic stories or complex statements.

---

[6]"Now, what's the difference between an invisible, incorporeal, floating dragon who spits heatless fire and no dragon at all?" (Sagan, *The Demon-Haunted World: Science as a Candle in the Dark*, p. 170–88)

The question is which view of the world we *should* choose. The "should" implies that this is ultimately a question of ethics. Obviously, if we evaluated theories independent of their complexity, our mind would be powerless as we would have to look at every possible theory and could not simply stop because its explanations are too convoluted or even infinite.

Occam's Razor is not a property of nature that somehow prefers simple solutions over complex ones. Occam's Razor does not even require that nature needs to be understandable. It is more a rule of thumb that, so far, has served us well in science and protected us from complex theories in instances when nature was, in actuality, much simpler. It expresses more the idea that when we look at nature, we first assume that its phenomena are complex, when in reality, the complexity simply stems from a lot of intertwined processes. Or to look at it from the other side: extraordinary statements require extraordinary evidence. And last but not least, simpler theories are simply easier to test and falsify if necessary.

Example

In the Middle Ages, people began to question the strange path of planets on the night sky. Mars is in "retrograde" every two years, following a strange circle (from the viewpoint of an observer on Earth, see Figure 3.1). While the explanation that Earth is simply the center of the universe (geocentrism) sounded like the most obvious idea at the time—from our perspective, the Earth does not move, but everything else does—the resulting diagram was anything but simple. It seems random. Drawing all the planets with the Earth at the center results in this flower-like diagram (Figure 3.2).

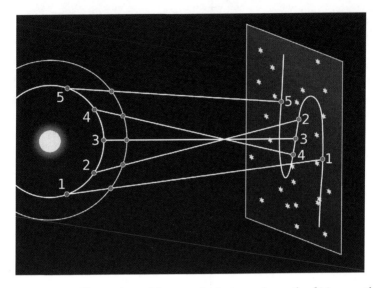

Figure 3.1: An illustration of the seemingly irregular path of Mars on the night sky from the point of view of Earth (image source: Brian Brondel, 2007).

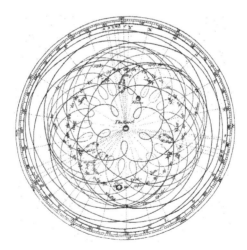

Figure 3.2: An illustration of the paths of the planets of the solar system with a geocentric world view (Encyclopaedia Britannica, 1st Edition, 1771).

This overly complex view of the solar system persisted until the Middle Ages. It was not an issue of technology; in Ptolemy's notes of the year 140 you can find thousands of star and planetary data and geometric ideas. The problem was more about the ability to read, the mathematical knowledge to read the notes, and ultimately the lack of availability of the notes.

Surprisingly, what ultimately led to a greater understanding of the universe were not telescopes, but the printing press. Only with the publication of Ptolemy's notes, in *Almagest*, the data found their way into the hands of people like Nicolaus Copernicus. Nearly 1,500 years after the data were first collected, Copernicus solved the problem of the complexity of the planetary orbits with heliocentrism—the idea that the Sun, not the Earth, is the center of the solar system.

Figure 3.3: An illustration of the heliocentric world view, with the sun, not Earth, being the center of the solar system.

While both views of the world, geocentrism and heliocentrism, are mathematically correct because they accurately predict the orbits of the planets, the system of Copernicus was much easier to handle. It added a layer of abstraction by looking at all the planets as separate entities instead of seeing the universe as an inexplicable series of snapshots, with the only task of science being to describe and to catalog instead of to *understand*.

Biography—**Nicolaus Copernicus**

Nicolaus Copernicus (1473 - 1543) was a Polish astronomer and polymath. The idea of heliocentrism—the Sun at the center of the solar system as opposed to the Earth being at the center of the universe—reached a wider audience with Copernicus' publication of *De revolutionibus orbium coelestium* in 1543. It triggered the Copernican Revolution which was a significant factor in the launch of a new age: the *Scientific Revolution*.

3.2.1 Addendum: The Hollow Earth

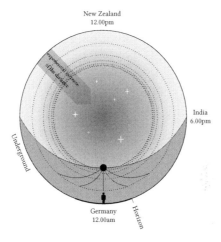

Figure 3.4: An illustration of the concave Hollow Earth hypothesis with the space being compressed in a way that the whole universe is in an infinitely small point in the center (image inspired by Joshua Cesa, 2010).

To get a better feeling for creating a hypothesis, let us look at the so-called "Hollow Earth hypothesis." This is a pseudo-scientific theory that we are actually living inside the Earth and not outside, with a "miniature sun" and a "star sphere" at its "center," and with a staunched space-time structure so that the distances to the sun and the stars are equal to our observed distances (see Figure 3.4). If we moved toward the center, we would become apparently "smaller" and "slower." But just as Copernicus asked how the geocentric model would look if you changed the point of reference from Earth to the sun, you could wonder how the model of the "Hollow Earth" hypothesis would look if you turned the space-time structure inside out, moving from "skycentric" back to a geocentric or heliocentric view of the universe.

Interestingly, you would end up with the heliocentric model! So, at its core, the Hollow Earth hypothesis is a "correct" model of the world, the physical laws would work in the same way, but the staunching of the space-time structure would complicate and obfuscate everything. And this is the moment for Occam's Razor, simply pointing out that including this staunching does not add anything and needs to be discarded.

When applying Occam's Razor, you have to keep in mind, though, that you first have to be aware of the complexity of the system in question and also to place it in some sort of context. It is not enough to know the (apparently) simpler solution, you have to know the whole context in order to decide what "simpler" in terms of Occam's Razor actually means.

Example

Imagine a time traveler from the year 1000 who has the chance to visit our world today. He will believe that something like the mobile phone is the product of a singular genius rather than an array of corporations around the world, exchanging goods and services required for the phone based on free, mutual trade. From his limited view, international cooperation of millions of people is less believable than magic.

Idea

For any given data, there can always be different models. Occam's Razor is useful when deciding which of the models is preferable by ranking them by their complexity.

3.3 Popular Science

> **Question**
>
> Science has become synonymous for "truthful." This has led
> people to try to emulate science without following the scien-
> tific method. Some have even started calling themselves "sci-
> entists" and used their titles to give their statements more
> meaning. Likewise, expressions such as "pseudoscience"
> have become an insult. What is the role of the word "science"
> in daily life?

> For a successful technology, reality must take precedence
> over public relations, for nature cannot be fooled.

—Richard Feynman, *What Do You Care What Other People Think?
Further Adventures of a Curious Character*

ARGUMENT FROM AUTHORITY · Just because there is an authority, be it a priest, politician, or scientist, who makes a statement, this does say anything about how truthful that statement
is. While the position or title of a person certainly means that
the person has been tested and has something to lose, all new
statements have to be proven. Usually, a title means nothing but
the proof that someone was serious about getting it, and is serious about keeping it. The quality of a title then ultimately depends not on its length, but on the community behind the title
that checks its quality. Experts can lie or make honest mistakes;
relying on someone's title does not protect you from that (cf. Cialdini, *Influence—The Psychology of Persuasion*, p. 208–36).

> **Example**
>
> The "cliff-jumper" argument involves the notion that we can only judge something if we have experienced it ourselves. If that were the case, we would first have to jump from a particular cliff in order to be able to argue for or against jumping from that cliff. This view originates from extreme empiricism, according to which there are supposedly no general principles in nature, but only statements applicable to a given situation. For an extreme empiricist, each new jump from a cliff would always be a new unknown—we could not draw conclusions from past observations that we could apply to the future or other similar situations.

The *argument from authority* is also used as a variant of the "cliff-jumper" argument: people are excluded from discussing a subject simply because they do not have the requisite title. Only a "true" scientist may provide his or her opinion about a subject. Only someone who "walked the walk" would be qualified. The fact of the matter is, though, more and more problems require interdisciplinary approaches. So, in fact, there are no "experts." What it comes down to is that there is no shortcut in science: you have learned the principles of the scientific method and of epistemology, and you have the ability to read and study papers. Relying on others (the so-called "experts") speeds things up, but also comes at a price.

Where I agree with the *argument from authority* is when it comes to the basic rules. You need to have a basic understanding of the scientific method, otherwise you should not be allowed to "play the game." Because in a way, science is a game with very strict rules. You can do research, you can publish articles, you can voice your opinions, etc. But you can only call your method "scientific" when you follow the rules. It is not enough to follow just some aspects of it (that would be cargo cult science or pseudoscience).[7] Sitting in an office, writing formally correct papers, and having a title does not make you "more scientific" if you do not test your ideas against reality in experiments.

> **CARGO CULT** · A *cargo cult* refers to the behavior where someone tries to imitate certain aspects of another (successful) person, expecting the same success. For example, celebrities are often on TV but just by managing to get yourself on TV, you will not necessarily become a celebrity.

The challenge of "being scientific" is that even not following the scientific method can yield usable results. And if that happens, this is sometimes being used as a justification to call the process that led to these results "scientific." But even if the process itself is formally correct, some of the studies are simply impossible to repeat and thus cannot be peer reviewed.

Example

Psychology obviously produces usable results. But strictly speaking, most papers on psychology are pseudoscience as they cannot provide a repeatable experiment because every single mind is unique. These experiments can, though, make way for the creation of a new hypothesis about the mind. Studies might show a difference in behavior between men and women. But with that result, you have only a tendency and you cannot make a statement about a specific person.

Until science and technology reach a point where repeatable experiments for high-level psychological investigations can be formulated, we have to base the conclusions on probabilities, tendencies, and generalizations. We might still produce usable results, but they will

---

[7]See *Philosophy for Heroes: Knowledge* for a discussion of cargo cults.

leave a lot of room for speculation and interpretation. As scientists in these fields, we have to stress that the scientific method was not employed with the same rigor as in the physical sciences. We have to acknowledge this limitation if we do not want to do a disservice to the whole field of science.

3.3.1 Communicating Science

A common problem is how to communicate scientific findings to the public. Every single writer—scientifically literate or not—can easily compete with any scientist. A scientist needs to find a network of readers and listeners. Academia provides this network to interested students and scientists, but not to the general public. The challenge is that academic funding of science only helps on the supply side, it does not build networks. Compare that to privately funded projects (including studies, books, etc.) where half of the work goes into creating a community and listening to the demands of the consumer.

Did you know?

There is of course also the other side, namely pseudosciences trying to co-opt the scientific community's efforts by offering "alternative" solutions, especially in fields where science does not yet provide "preferable" outcomes (be it in medicine, or be it in philosophy). Simply producing something is not enough, you also have to create networks of interested readers. If honest scientists evade this responsibility, "alternate science" will sell its ideas—with exaggerations or even falsehoods added. So, for scientists to really affect change, they need to learn to lead by example and tell a compelling story.

⟶ Read more in *Philosophy for Heroes: Epos*

Of course, the market fills that gap. Unfortunately, when information is to be made available for free on websites, the price the visitor has to pay is marketing. Only articles with an intriguing title will get enough readers to pay for the investment. Tracing back articles from popular science magazines or websites to their actual studies usually leads from "Sleep less, live longer, scientists say in new study" to "Researchers discovered that all other factors of diet, age, health and lifestyle being equal, those who slept eight hours a night were 12 per cent more likely to die within a six-year period than those who slept seven hours" to "Additional studies are needed to determine if setting your alarm clock earlier will actually improve your health."[8] In systems as complex as the human mind or body, it is relatively easy to find correlations when asking a large enough group of people about their behavior. But these are always just correlations, not causations. That is why most studies contain the mandatory sentence about "additional studies needed."

Idea

The troubling issue of modern science is that some scientists see themselves no longer as part of a branch of philosophy, but as philosophers themselves. As a leader, it is not enough to just remain in one field of study, trying to subjugate all other fields of study. Your task is to reach out to other fields and unite them, to go back and forth between the "what is" and "how do we know."

---

[8] Abraham, *Sleep less, live longer, scientists say in new study.*

3.4 Heisenberg's Uncertainty Principle

 You ask me if an ordinary person could ever get to be able to imagine these things like I imagine them. Of course! I was an ordinary person who studied hard. There are no miracle people. It happens they get interested in this thing and they learn all this stuff, but they're just people. There's no talent, no special ability to understand quantum mechanics, or to imagine electromagnetic fields, that comes without practice and reading and learning and study. I was not born understanding quantum mechanics—I still don't understand quantum mechanics! I was born not knowing things were made out of atoms, and not being able to visualize, therefore, when I saw the bottle of milk that I was sucking, that it was a dynamic bunch of balls bouncing around. I had to learn that just like anybody else. So if you take an ordinary person who is willing to devote a great deal of time and work and thinking and mathematics, then he's become a scientist!

—Richard Feynman, *Fun to Imagine*

Please note: this and the following sections are but an introduction into quantum mechanics—the current field of research in physics explaining the world at the particle level. In later chapters, we will return to the theory to get a deeper understanding, just like this book goes back to Philosophy for Heroes: Knowledge. My own attempt to understand quantum mechanics took me more than two years, so please be patient —it is by its very definition a non-intuitive topic! I also recommend reading the referenced secondary literature to really wrap your mind around it.

Question

How can we actually measure the position of a particle?

Heisenberg's Uncertainty Principle was formulated in 1927 by Werner Heisenberg. It states that you cannot determine the location as well as the impulse (the energy of the movement) of an entity with infinite precision. An interpretation of this principle is that there cannot be objective measurements insofar as measurements always influence the entity that one is measuring. This is especially an issue for smaller particles like electrons.

To understand this principle, it is important to understand how we perceive the world. A measurement is nothing but a form of perception or observation. In contrast to the ancient view that our eyes send out "seeing rays," we do not perceive our environment directly but rely on light reflected from objects. In order for an object to reflect light, light has to be beamed at the object. When examining very small particles, we encounter the problem that the smaller the particle, the smaller the wavelength of light has to be in order to be reflected at all. But the smaller the wavelength of light, the more energy is required and the more the beam of light influences the particle.

We can determine the position of a particle with precision limited only by the technology currently available to us. But this determination could cause the particle to be slowed down or redirected in its course, destroying the information about the impulse of the particle. Likewise, one could measure the impulse of a particle by simply having it hit a screen; however, this destroys the information about the position of the particle (replacing it with the known position of the screen).

> **Example**
>
> Sound waves are the directed vibration of air molecules
> caused by a sound source. There is no way to determine
> the frequency or other properties of a sound wave by a sin-
> gle "snapshot" of the air. Even if you knew all the positions
> of the air molecules at a certain point in time, you would
> know nothing about how they vibrate. If you instead used
> a microphone and let the air molecules hit a membrane, you
> could get information about the whole sound wave and its
> frequency.

3.4.1 Revisiting Objective Perception

With this new knowledge about the inner workings of the physical
world, it is time to reflect on our existing concepts. This reflection is
not a violation of our philosophic principles established in the first
place, quite the opposite: we refine our epistemology and ontology
constantly in order to get a better view of reality.

So, what follows from our inability to measure location and impulse
of entities with arbitrary precision? For the creation of concepts,
this does not bother us because we omit the measurement anyways.
We have to take a step back, though, from the idea that by building
better measurement tools we could measure anything. The underlying issue simply is that we are part of the universe, so any action we
take—including observing it by measurements—influences the universe. The consequence is that we cannot be omniscient. But as
we have established in the first book, omniscience is not required
in order to have an objective perception of reality (and vice versa,
objective perception does not mean potential omniscience).

Something seems wrong, though. Can we really say that we know the properties of a particle if we cannot determine their position and impulse at the same time? Are entities like electrons or atoms really entities in the classical sense, similar to tiny billiard balls? Taking a look at atoms, we really cannot take a "photo" of an atom with its electrons, yet a very popular depiction is the core being surrounded by electrons (see Figure 3.5):

Figure 3.5: A common depiction of an atom with distinct electrons as "billiard balls" (image source: shutterstock).

But this depiction does not reflect reality. Classical mechanics, with an entity-based philosophy, with a single particle (a negatively charged electron) orbiting the atom core (neutrons and positively charged protons), provides no explanation of why the electron does not fall into the core.

In classical physics, there are no fixed *quanta*. In quantum mechanics, there are no distinct electron particles but electron clouds with certain energy states around the nucleus based on the probability of where the electron could be.

In that regard, Heisenberg's Uncertainty Principle correctly identifies that position and momentum cannot be measured together. But by talking about measuring position and measuring momentum, it could falsely imply that particles are in fact particles in the classical sense with a position and momentum and that only by some weird circumstance, we are unable to measure both together.

How would we approach this issue with philosophy? The problem the uncertainty principle poses for our definitions is that concept creation involves making observations of reality, and omitting measurements in order to focus on the actual properties of an entity. But if we cannot make those measurements in the first place, they cannot be omitted either! Logically, it would follow that we cannot create concepts of particles in the (small) quantum world that do not have both a position and momentum. But if we cannot create a concept based on entities for those small particles, what would that mean for our ontology?

Let us sum up what we have so far:

1. Concept creation means to make observations and omit measurements.

2. The uncertainty principle states that for small particles, objective measurements cannot be taken.

3. If no measurements can be taken, they cannot be omitted.

4. From (1) and (3) follows that we cannot create a concept for these small particles.

> **Idea**
>
> Particles are in fact not tiny (classical) billiard balls. Measuring a particle's position is only possible by influencing the particle, making it impossible to know its original position.

Just as in the Middle Ages, when people learned that our eyes do not send out "seeing rays" and had to re-examine their epistemological premises, with the discovery of quantum mechanics, we likewise need to go back to re-examine our ontological premises. In *Philosophy for Heroes: Knowledge*, we started with looking at the world as consisting of entities. Our journey through science has now brought us to the very edge of this entity-based thinking, Western philosophy, and classical mechanics. Having established that the world does not consist of tiny billiard balls, we now have to figure out what "particles" actually are. This will be the story of the following pages.

3.5 Development of Quantum Mechanics

> **Question**
>
> Is light a wave or a particle?

> [...] when we allow for the dynamics of gravity and quan-
> tum mechanics, we find that [our] commonsense notion is no
> longer true. This is the beauty of science, and it should not be
> threatening. Science simply forces us to revise what is sensi-
> ble to accommodate the universe, rather than vice versa.

> —Lawrence Krauss, *A Universe From Nothing*

Quantum mechanics could just as well be called "wave mechanics."[9]
That is to stress the fact that classical mechanics dealt with the universe as comprised of particles as if they were small billiard balls,
while quantum mechanics sees reality as particles acting as waves,
not as billiard balls. Once we group particles together as larger, visible objects, the results from classical "billiard" mechanics are good
enough to explain most phenomena.

But let us first define what we mean by "classical mechanics." It
refers to physics before the 20th century. Back then, scientists
thought that they were very close to a complete description of nature with everything being particles. Likewise, it refers to physics
at low speeds (nowhere near light-speed). In 1802 (through Thomas
Young's double-slit experiment), it was discovered, though, that
light acts like a wave rather than like individual particles. Light
cast through slits or holes results in an interference pattern (see Figure 3.6).

---
[9]cf. Bell, *Speakable and Unspeakable in Quantum Mechanics*, p. 187.

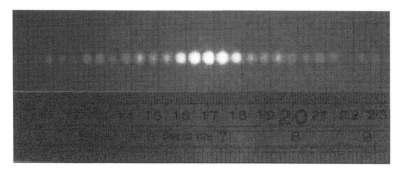

Figure 3.6: A diffraction pattern of a 633 nm laser with 1mW power through a grid of 150 slits, 0.0625 mm each, 0.25 mm separations between their centers (Shim'on and Slava Rybka, Ben Gurion University Physics Department).

Example

You can easily recreate the double-slit experiment with a laser pointer wrapped in aluminum foil. Use a needle to puncture one hole into the aluminum foil and compare the light pattern with the one after having punctured a second hole into the aluminum foil. You will see an interference pattern similar to that of water waves: if you throw two stones into the water, the hills and valleys of the waves overlap.

Generally, water waves are a good model for the interference pattern of light. On the following pages, you can see real world examples and schematics illustrating the properties of waves (see Figures 3.7, 3.8, and 3.9).

Figure 3.7: Diffraction waves seen along the coastline near Pesaro (image source: shutterstock).

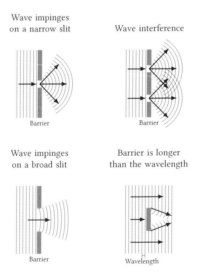

Figure 3.8: Different behaviors of waves when facing an obstacle (image source: shutterstock).

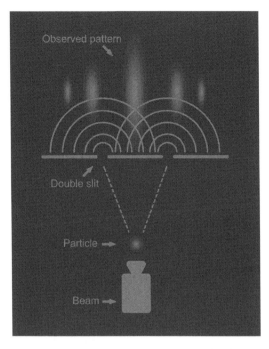

Figure 3.9: The two-slit experiment showing a diffraction pattern where the light waves interfere with each other (image source: shutterstock).

The idea that light is a wave was cast into question later when examining thermal radiation, the light that a heated object emits. Again, something you can observe yourself when heating up an oven and watching how the coils turn from black (infrared), to red, to white. If you had tools to objectively record the temperature and the wavelengths and intensity of the light, you would see how the intensity of the light first rises, and then sharply drops off for wavelengths below ultraviolet light (see Figure 3.10):

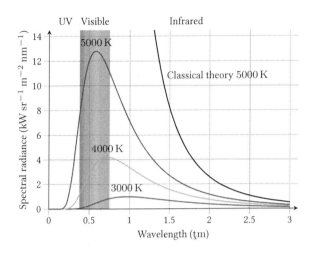

Figure 3.10: A comparison of the prediction of black body radiation according to the classical theory with actual measurements.

How did the scientists come up with a theory for thermal radiation using classical mechanics? The task was to calculate the amount of energy released by an object depending on the wavelength. The simple approach was to take a cube and calculate how many waves of a certain wavelength fit into it. This showed accurate results for larger wavelengths, but catastrophically failed for smaller wavelengths. Once the wavelength tends toward zero, the supposed emitted amount of energy by the object would tend toward infinity.

The solution was to go back to the original idea: light has properties of waves *and particles*. The idea was that light comes in packages ("quanta"—the plural of quantum which is Latin for "small amount"). Heat causes electrons of an atom to move to a higher level of energy. When the electrons return to their normal level, photons are generated and emitted as an electromagnetic wave.

But these different levels of energy are not continuous. Instead, you can imagine them like distinct electric neighborhoods around the atom. Either the energy is enough and they are pushed into the other neighborhood, or they do not move at all. The resulting photon has a fixed speed (light speed), no mass, and an energy dependent on its wavelength.

Example

Think of a vending machine that accepts only quarters and not pennies: no matter how many pennies you put in, it will never add up to a quarter. Likewise, imagine an on/off switch on the machine that you cannot push continuously, but either it does not move at all, or it moves all the way into one or the other position. This is what the energy states of electrons are like.

Example

The wavelength of a photon can be compared to sound: when you hit a drum, the membrane vibrates back and forth. This leads to the air being compressed and uncompressed in a regular frequency. This local oscillation is propagated throughout the room until it hits, for example, your eardrums, translating the oscillation of the air to a vibration of your membranes to nerve signals. Larger drums vibrate more slowly, leading to longer (deeper sounding) wavelengths. As opposed to moving air molecules, the photon vibrates the electric field. These vibrations can likewise cause cells in your eyes to be activated, translating them into nerve signals as well. Electromagnetic waves with larger wavelengths are called infrared, radar, microwave, or radio waves.

Idea

Light has properties of both waves and particles. It causes interference patterns like waves, but also comes in packages ("quanta"), hence the name "quantum theory" was born.

On top of quantum mechanics, relativistic mechanics was developed in 1905 with the publication of Einstein's paper on special relativity (*Zur Elektrodynamik bewegter Körper*). Putting everything together, we can distinguish four major fields in physics distinguished by energy and speed (see Figure 3.11): classical mechanics (low speed, high energy), relativistic mechanics (high speed, high energy), quantum mechanics (low speed, low energy), and quantum field theory (high speed, low energy).

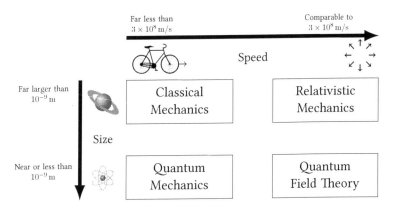

Figure 3.11: Simplified illustration of the four fields of physics distinguished by energy and speed (image inspired by Yassine Mrabet, 2008).

3.5.1 Quantum Computing

Question

How can a quantum computer be significantly faster than a conventional digital computer?

A real-world application for the quantum theory is quantum computing. This field of science was initiated by the work of Paul Benioff and Yuri Manin in 1980, Richard Feynman in 1982, and David Deutsch in 1985. To understand quantum computing, let us first examine classical (digital) computing. Classical computing is based on binary digits (bits): either there is a current, or there is no current. With this basic distinction, logical "gates" work. Gates are small electrical relays with input and output wires.

Example

In classical computers, the logical AND gate has two input wires and one output wire. When there is current on both input wires, the AND gate creates current on the output wire. If there is no current or current on only one of the wires, there is no output current. Similarly, OR, XOR, NXOR, NOT, etc. gates can be configured, with different input and output configurations. With these simple mechanisms, computers can be built.

Qubit · A *qubit* can be compared to a wave (as opposed to a discrete unit as with digital bits). Combining several qubits with a quantum gate results in an instant calculation of an interference pattern between the waves, which can be used to significantly speed up processing.

Now, what about quantum computing? Quantum computers also have gates and "bits," but both are different than the classical electrical bits and gates. Hence, they are called "quantum bits" (qubits) and "quantum gates." To understand how those quantum gates work, we need to remember what we have said about quantum mechanics: quantum mechanics is basically wave mechanics. And a quantum computer is basically a "wave computer." How can we build a "wave computer"? For that, let us go back to one of our examples: water waves and the double-slit experiment. Like the electrical AND gate, the double-slit setup has two inputs (the two slits) and one output (the screen). If we send no wave (or close both slits), the screen will show no pattern, the water is quiet. If we close either one of the slits, a single wave hits the screen. If we open both slits, we will see an interference pattern. So, already, we have some kind of "gate" here: an interference pattern will show up only if we send waves through both slits. There exists only "no wave" (0) or "wave" (1), and waves you send through the slits can be of any configuration between 0 and 1. Now, imagine a 10-slit setup: already, this configuration "calculates" the interference pattern of 1,024 possible different inputs! Depending on whether you are sending waves through individual slits or not, a different interference pattern emerges.

Currently, a number of companies and organizations are trying to build larger quantum computers. Depending on what counts as a quantum computer, the ranges are between 50 and 1,000 qubits. With so many qubits, this computer would be much, much faster than a classical computer. Again, not because it does more operations per second, but because a single operation instantly calculates the interference pattern of all qubits. And more qubits do not just increase the speed in a linear fashion. Doubling the number of qubits does not simply double the speed but leads to a multi-fold increase of speed. The larger the problem, the faster such a quantum computer would be compared to a classical computer.

The challenge of actually *building* such a quantum computer lies in quantum mechanics: in order for those quantum gates to work, they have to be cooled down to (near) absolute zero and any electromagnetic radiation needs to be shielded completely. Connecting all those qubits in a meaningful way and bridging the results into the "classical" world are additional challenges.

It might take another decade before true quantum computers become generally available, though, because error correction and reliability are still significant issues. Ultimately, a quantum computer ditches the idea of having a digital near-100% accurate result[10] in favor of speed: it gives you the option if you can live with it—quick calculation with the possibility of false positives might sometimes be more important than 100% accuracy.

Idea

Quantum computing is so much more powerful than conventional computing because the way the waves interfere with each other is "computed" by nature for free. Basically, quantum computers repeatedly run small physics experiments, which makes solving problems of a similar nature very efficient. The difference, compared to a conventional computer, is that the interference pattern emerges in a few steps depending on the accuracy you want. Thus, the calculations of a quantum computer are much less problem-size dependent.

---

[10]The actual source of computation errors in digital computing is the memory, not the processor.

3.6 Copenhagen Interpretation

The unexplained mystery of thermal radiation (among others, like the photoelectric effect) led to the theory that light comes in packages: the quantum theory. The first interpretation to explain the quantum theory epistemologically was the *Copenhagen interpretation*. It was first formulated in 1927 by Niels Bohr, John von Neumann, and Werner Heisenberg, and became the most popular explanation of what is actually going on at the quantum level.

It stated that particles do not exist (meaning that they do not have definite properties) as long as you do not measure (observe) them. The "observation" part does not depend on whether a human scientist is actually looking at the particles; a measuring device interacting with a particle is enough. Before the measurement is taken, a particle exists only as a probability, e.g., 50% outcome A / 50% outcome B. But after the measurement, is it reduced to a single value. For example, take the process of flipping a coin: while it is in the air, you cannot tell on which side it will land. Only once you "take the measurement," i.e., catch the coin, it is reduced to heads or tails. In the *Copenhagen interpretation*, this process is called the "collapse of the wave function."

With that in mind, let us look again at a common experiment, the already mentioned "double-slit experiment." You have a screen on the right side, and a screen in the middle with either one or two slits (see Figure 3.12). In classical mechanics, if you shoot small particles of matter on a screen with one slit, a simple probability distribution shows up on the screen on the right side: most of the particles will hit the middle. With two slits, you end up with two peak locations on the right screen. On the other hand, if you were to send a water wave through the slits, two separate waves would come through, each interacting with the other, resulting in an interference pattern, where parts of the wave reinforce each other, and parts cancel each other.

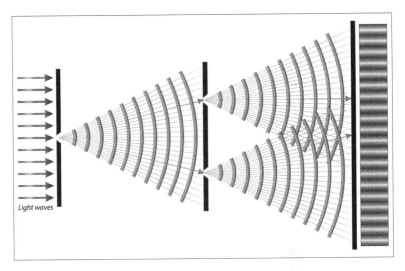

Figure 3.12: Light passed through two slits cause an interference pattern on the screen.

In quantum mechanics, this kind of interference pattern also shows up with particles. Even if you send individual particles one after the other (as opposed to a continuous stream of particles), an interference pattern will emerge. With only classical mechanics in mind, it looks as if the particle passes through both slits and then interferes with itself: it acts like a wave. This wave behaves differently from classical waves as you will ultimately see only one outcome (and not a "wave" hitting everything in its path).

In this context, we have to remember the previous section about *Heisenberg's uncertainty principle*. We cannot observe something without somehow interfering with a particle. For our eyes (or a camera) to work, we need to cast light waves on an entity. This means that adding any type of detector ("observer") near the slits to check through which slit the particle passed will not help: this will collapse the wave function and the wave will return to acting like a particle.

But what exactly is an "observer"?

3.6.1 The Observer

In classical science, we look at the world from the outside. Entities and events of every kind are dissected into their parts and then separately surveyed and categorized. This includes the observing scientist and the test equipment: in order to create an observer-independent model of reality, the result of the test must not be influenced by the test itself. This also makes it possible for the experiment to be repeated by other scientists.

Example

When measuring the heat generated by a chemical reaction, the temperature first has to be controlled by putting the experiment into a sealed container of constant temperature and leaving the scientist outside in order for the body temperature to not influence the result. This is at least the basic philosophy of classical physics. This approach has brought us as far as the moon—at least the moon landing could have been successful with classical mechanics only.

> For, as this interpretation now stands, it is always necessary to assume an observer (or his proxy in the form of an instrument) which is not contained in the theory itself. If this theory is intended to apply cosmologically, it is evidently necessary that we should not, from the very outset, assume essential elements that are not capable of being included in the theory.

—David Bohm, *The Undivided Universe*

The challenge is that if you focus only on the "how" (epistemology) and not on the "what is" (ontology), you end up being a disembodied observer. But in fact, there is no such thing as a disembodied observer, as you are always part of any experiment you are doing. At the small scale, you cannot observe a particle "from the outside" without influencing its state, no matter what precautions you take. As we have concluded in the first book, epistemology and ontology are strongly intertwined! You cannot start with epistemology and then create an ontology out of it and vice versa.

To repeat a quote we have read in *Philosophy for Heroes: Knowledge*:

> [Ontology] and epistemology are simultaneous—what exists and how we know it are the foundation that starts together. And that's why the very first axiom is "Existence exists, and the act of grasping this implies there is something, and we have the faculty for being aware of it." And thereafter we shift back and forth, "We have consciousness," "A is A," "Existence is independent of consciousness," "We acquire knowledge by reason," and so on. [Ontology and epistemology] are completely intertwined.

—Leonard Peikoff, *Understanding Objectivism*

Many technologies of the last century—whenever accuracy was needed—be it the atomic clocks used for the Global Positioning System (GPS), lasers, semiconductors for computers, or Magnetic Resonance Imaging (MRI), rely on quantum mechanics (and the theory of relativity). There, you are dealing with very low energy levels requiring high accuracy. When experimenting with individual particles, the influence of the observer becomes significant. As discussed in the section about *Heisenberg's uncertainty principle*, any measurement of these particles would influence their impulse or location.

In the first book, *Philosophy for Heroes: Knowledge*, we stressed the important difference between the approach in Objectivism and in (classical) science. In Objectivism, you start out with your own existence to make a statement about the world. Its basic axiom is "I exist, I have an identity, and I can be conscious about this fact." So, the essence of philosophy is to understand yourself as being part of reality, rather than isolating yourself from reality as a passive observer. Hence, at least on the epistemological level, our philosophy is in line with modern physics.

But having confirmed the quantum theory with our philosophy is not enough. The theory only provides an epistemological explanation.

What is really going on?

Quantum theory is primarily directed towards epistemology which is the study that focuses on the question of how we obtain our knowledge [...] It follows from this that quantum mechanics can say little or nothing about reality itself. In philosophical terminology, it does not give what can be called an ontology for a quantum system.

—David Bohm, *The Undivided Universe*

Example

The idea of the Copenhagen interpretation that nature stops and sits on her hands when a measurement is made is strange, ontologically speaking. Imagine a lake. There is someone throwing stones into the water, causing lots of ripples. Now, close your eyes. You can no longer see the ripples in the water, but you assume that they are there. Then, you hold your hand in the water to feel the ripples splashing against your skin. Suddenly, aside from the water splashing against your arm, the whole lake becomes quiet. You have "collapsed the wave function" of the lake into "real water particles hitting your hand" by taking a measurement. You remove your hand from the water, and open your eyes: the lake is again full of ripples.

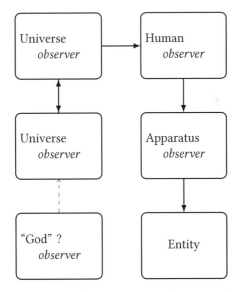

Figure 3.13: Illustration of the problem of the original observer in the Copenhagen interpretation: in order to observe an entity, the observer itself has to be observed, resulting in an infinite recursion.

Taken to its conclusion, the Copenhagen interpretation posits that nature is undefined until it is observed. That poses a problem: what about the observer? An undefined observer cannot observe until another observer observes the observer. An undefined entity has no properties! Ultimately, this adds an infinite recursion (see Figure 3.13) to the answer—with no explanation who the "first observer" would have been. It is then argued that this would be proof that the universe has some sort of supernatural observer (e.g., God) that observes everything in order for it to come into existence.

> [...] from some popular presentations the general public could get the impression that the very existence of the cosmos depends on our being here to observe the observables. I do not know that this is wrong. I am inclined to hope that we are indeed that important. But I see no evidence that it is so in the success of contemporary quantum theory. [...] The only 'observer' which is essential in orthodox practical quantum theory is the inanimate apparatus which amplifies microscopic events to macroscopic consequences.

—John Stewart Bell, *Speakable and Unspeakable in Quantum Mechanics*

Similarly, physicist Erwin Schrödinger saw a problem with the *Copenhagen interpretation.* To illustrate the problem, he came up with the thought experiment "Schrödinger's cat." In this experiment, there is a decaying nuclear isotope connected with a detector which releases a deadly gas into a box that contains a cat (see Figure 3.14). With the ontological interpretation of the Copenhagen interpretation, the cat would be both dead and alive: just like the particle, it would have no definite properties; the cat would be the macroscopic representation of the (supposedly) undefined state of the microscopic particle.

Figure 3.14: Schrödinger's cat, both dead and alive (image source: shutterstock).

> [...] the *Copenhagen interpretation* as such totally evades the real question, which is how the quantum-mechanical description at a microscopic level becomes converted into a classical one at the macroscopic level [...]

—David Bohm, *Quantum Implications: Essays in Honour of David Bohm*

The crux is the detector that connects the microscopic with the macroscopic world. It is as if you held a pin that is so pointed that there is but a single atom at its end, touching another particle. In that regard, your hand holding the pin is "classical," the particle is "quantum mechanical," and the pin is something in between. This "something in between" is the very point of the example, showing that the concept of splitting both "worlds" does not make sense.

> How exactly is the world to be divided into speakable apparatus [...] that we can talk about [...] and unspeakable quantum system that we cannot talk about? How many electrons, or atoms, or molecules, make an "apparatus?"

—John Stewart Bell, *Speakable and Unspeakable in Quantum Mechanics*

QUANTUM WEIRDNESS · The concept of *quantum weirdness* refers to the unintuitive results we see when looking at effects in the quantum world. Intuitively, we expect everything to act the way it does in our immediate, slow-moving and high-energy "macro-world." But for particles, this intuitive approach does not work, hence we call the quantum world "weird" in that regard.

Because there is no clear line between the microscopic and macroscopic world, the *Copenhagen interpretation* of ontology needs to be cast into question. While this approach of seeing the observer and observed as two separate worlds is very problematic, this of course has no influence on the quantum theory, i.e., the mathematical calculations of how particles behave at a quantum level. But it has everything to do with the ontological *interpretation* of these results. And as we will see in the following pages, there are other interpretations based on the same mathematical calculations that do not introduce such "weirdness."

Idea

In the ontology of the Copenhagen interpretation, the world is based on pure mathematics. Physicists stay away from the question where these probabilities originate. As the Copenhagen interpretation is based on the (mathematically correct) quantum theory, its predictions are the same as any other interpretation—but it cannot *explain* anything.

3.7 Interpretations of the Quantum Theory

3.7.1 Many Worlds Interpretation

Question

Why could the many worlds interpretation be called ontologically wasteful?

An approach to explain what is going on at the ontological level is the so-called *many worlds interpretation*. Instead of claiming that until something is measured, it is not defined, it states that at all times, we have a definite history. This means that Schrödinger's cat is always either alive or dead, no matter whether you open the box or leave it closed. In *that* regard, the *many worlds interpretation* is deterministic. There is a clear path that leads from a past situation A to the current situation B.

But in order to be deterministic, it needs to do away with another property of classical mechanics, namely that there is but one reality. One needs to assume that with every measurement (interaction between particles), the universe splits into an infinite number of copies, each copy representing one possible outcome of the probability of the measurement. Looking toward the past, you always have but one "parent universe." Looking toward the future, you always have an infinite number of universes. In the case of Schrödinger's cat, you can imagine that the cat stays alive in at least one branch of this tree of possibilities. While it becomes deterministic by using this "trick," you still do not learn anything about how the universe *actually* works (there are no cause-action pairs). Ontologically, the universe would be still based on probabilities—you do not know *why* your universe "decided" to pick a certain branch.

When talking about universes, we have talked about the infinite universe, as well as the idea of having an infinite number of universes. *The* universe is a large (and possibly infinite) canvas, with big bangs at random places. If we could look at this canvas from afar from outside the big bangs, we would see independent "pocket universes" here and there, like stars in the night sky. You could theoretically travel from one pocket universe to the other if you traveled long enough. But the *many worlds interpretation* refers to a *different* type of "infinite" or "many." The branching in the *many worlds interpretation* does not refer to those pocket universes, it means whole worlds, which in principle (and not because they are so far away) could not interact with each other. So, to summarize:

- **Universe:** the infinite canvas on which big bangs (creating pocket universes) happen.

- **Pocket universe:** created by big bangs and refers to the observable universe.

- **Many worlds:** distinct universes which, in principle, cannot interact with each other. They are created with every measurement.

While this approach sounds far-fetched, it is one of the many possible interpretations of the observations from experiments related to quantum theory. It relies on the same proven mathematical results we get when using other interpretations. What is obviously different is its ontology. It does assign a definite state to all entities at every point in time. But it is an extremely wasteful interpretation, violating the Occam's Razor principle to the largest extent: with each interaction on the quantum level, it creates an infinite number of copies of the universe. And with infinity, strangeness ensues: with all the possibilities becoming a reality in some universe, you will have one universe with a basically immortal cat, defying all odds. In contrast, with the *Copenhagen interpretation*, the survival rate of the cat would tend toward zero.

Idea

Instead of saying that the quantum world is based on probabilities, in the many worlds interpretation, it is argued that the universe is fully deterministic—insofar as it has a definite history—but that at each "moment," the universe splits into an infinite number of copies, each representing one choice among the probabilities. It would be ontologically wasteful as it requires the creation of an infinite number of universes. In addition, you would still not know *why* a certain branch was chosen.

3.7.2 Hidden Variables

Question

What sets the hidden variables interpretation apart from the Copenhagen interpretation of quantum mechanics?

In the *Copenhagen interpretation*, the state of the universe is "undefined" before it is measured. That is what "quantum weirdness" is largely about, precisely because that is not how probabilities usually work. "Schrödinger's cat" was all about highlighting the weirdness of the *Copenhagen interpretation*, by taking the supposedly undefined state at the quantum level to the macroscopic level: a cat that would be both dead and alive.

So, a different approach does away with the whole idea of probabilities. Instead of assuming that the collapse of the wave function is based on probabilities or chance (or branches off into infinite worlds), it assumes that there are "hidden variables." According to this idea, the process is not truly random, but we simply lack an insight into the hidden workings of the system we are measuring. For example, there is a $\frac{1}{6}$ chance to roll a 7 with a pair of dice, but there is a 100% "chance" for one result to show up when it follows from the previous state. If you knew the exact position of the dice as well as the vector of the dice throw, you could calculate the result.

Unfortunately, when using hidden variables to do away with the probabilities, we need to introduce a different kind of "weirdness": non-locality. Non-locality is the idea that an action at one place influences something at another immediately. But at its core, non-locality is a simple concept and not "weird" at all.

Figure 3.15: Two sides of a coin (image source: shutterstock).

Suppose I, without looking, take a coin and split it horizontally in two. Then, still with neither one of us looking, I give you one half and leave on a journey to the stars. The first person who looks at the coin immediately knows which part the other person has, despite being light-years away. There has been no "faster than light" communication, yet our knowledge about the other half of the coin was instantaneous over light-years.

Non-locality seems surprising only if you approach nature with the worldview of the *Copenhagen interpretation*, namely that nature waits until you take a look at her. If you assume that the coin halves had no definite property until they are measured, and that the fundamental properties of the world supposedly consist of mathematical probabilities that are only realized when being measured, you might find this "instant" communication between the coin halves as some sort of "spooky action at a distance" that seemingly violates the limits of the speed of light.

The *Bohm interpretation* sees the quantum world instead as particles guided by a "pilot wave." This wave has properties and constitutes what we previously have called the "hidden variable." Just like the halved coin, a wave is intrinsically non-local: you throw a stone into the water and the corresponding wave contains the information of the stone throw. No matter how far it travels, the information about the initial throw is carried to multiple places. In the case of the quantum level, it is "hidden"—maybe hard or impossible to measure, but it is (ontologically) there. The place where the stone hits the water could be compared to the moment where both coin-carriers part ways, and the "wave" is the person traveling away with half of the coin. Looking at one half of the coin tells you of course instantly something about the other half of the coin because both halves are part of the same wave ("entangled") which was created by a common "throw" (the division of the coin).

Idea

Scientists have found it difficult to understand the Copenhagen interpretation as it does not explain where the probabilities come from. The hidden variables interpretation replaces the probabilities with a fixed "hidden" cause, the so-called pilot wave. It might be hard or impossible to measure, but they are (ontologically) there. As such, the hidden variables interpretation is deterministic—as opposed to the Copenhagen interpretation.

3.7.3 Quantum Effects on a Macroscopic Scale

Figure 3.16: Quantum effects can be modeled with hovering water or silicone oil droplets (image source: shutterstock).

With the help of a vibrating bath of silicone oil and individual droplets, quantum effects could be simulated on a macroscopic scale (see Figure 3.16). The vibration of the bath allows droplets to hover over a packet of air just above the surface for a long time. Below the droplet, there are waves. A moving droplet will ride along its waves, moving in directions or seemingly randomly. Surprisingly, tracing these paths, a pattern emerges that is very similar to that of particles at the quantum level. When these droplets hit a double-slit experiment, they do not act like the previously mentioned billiard balls. Instead, they act like a wave (see Figure 3.17). The guiding wave they take with them while floating over the water bath influences their path through the slits, just like it was observed with particles at the quantum scale.

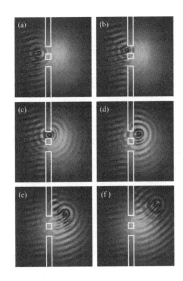

Figure 3.17: Using a vibrating bath of silicone oil, the results of the double-slit experiment can be recreated in the macroscopic world (image source *Double-slit experiment with single wave-driven particles and its relation to quantum mechanics*).

3.7.4 Outlook

Due to what we have said about Heisenberg's uncertainty principle, we cannot tell if a particle either does not have both impulse and a location and simply acts according to probabilities, or if there is some kind of "hidden variable" within a particle. But what is the "mechanism" within a particle? If we cannot peek into the particle and its workings, how can we tell if there is some random factor involved?

The major challenge in science is that when looking at the universe, we are always part of what we are looking at. We cannot put the universe under a microscope and look at the universe from the outside. It is us, the universe, that looks at itself. In physics, this problem has led to two approaches: description of the probabilities of how particles behave on the one hand, and, on the other, interpretation of the underlying reasons that these probabilities appear.

The first approach simply focuses on the math of everything related to this "jiggering" on the particle level. From that, the so-called "quantum theory" was born. It is one of the most tested theories in science and has never been proven wrong in the 100 years since its development. With it, you can compute the pattern that a stream of particles will take. This approach does not "peek into the box" but simply assumes that at a certain level, the world is ruled by probabilities, with no internal or external state influencing them.

The second approach deals with the *interpretations* of the quantum theory. Scientists want to know how it actually works, they want to *understand* it. In physics, this has led to a number of interpretations of the universe. In a way, it is another example for the list of theories we have discussed in the context of Occam's Razor: there are an infinite number of interpretations of a phenomena, but we want the one that does not add unnecessary baggage.

> ### Example
>
> We can look at the universe with the Earth as the center and all the planets and stars moving around it in strange paths. Or we simplify the model and have the Earth and planets orbiting the sun in nice ellipses. We can look at the Earth as a planet with a (more or less) planar space-time, or we look at the Earth as a hollow sphere with us living on the inside and with space-time stretched in a way that the whole universe is condensed in the center of the sphere.

All interpretations certainly require an adaption of our basic philosophy. But just because quantum mechanics is based on mathematics to make predictions, and just because many of today's technologies are based on the results of quantum mechanics, we do not have to accept all the weirdness that comes with the Copenhagen interpretation. It is certainly *determinism* that is most compatible with the philosophy outlined in the first book of the series, *Philosophy for Heroes: Knowledge*, and thus also the foundations of science.

The unanswered question for non-deterministic interpretations is how they can arrive at their conclusions while not at the same time undermining the philosophic requirements of the very scientific method they are using. Taken to their conclusion, for example the Copenhagen interpretation has to introduce either an arbitrary distinction between the micro and macro world, or an ultimate observer—a "God"—that observes the observers in order for them to exist. You end up with the same mathematical results, but again introduce a lot more ontological baggage as opposed to deterministic interpretations like we have laid out here.

Did you know?

In terms of the quantum theory, we face the problem that *all* interpretations introduce at least one non-intuitive element, be it indeterminism, non-locality, quantum consciousness, or infinite worlds. But all interpretations are "correct" insofar as they are based on the exact same observable results. So, for quantum mechanics, Occam's Razor is too blunt. Instead, it seems that individual psychological preferences rule which of the interpretations people will prefer. Our psychological make-up has an influence on our preference for certain epistemological and ontological views.

$\longrightarrow$ Read more in *Philosophy for Heroes: Act*

Either way, I hope that the discussion here has shown one central point: that we are dealing always with interpretations, not (testable) theories. They are philosophical explanations to explain laws of physics. At this point, no interpretation really has the upper hand and we should wait until more research has been done. At least the simple two-dimensional model from above seems to be a good candidate that could serve as a model to understand what is happening at the microscopic level. With this model and the deterministic *Bohm interpretation*, it looks like science is getting closer to a reasonable model for quantum mechanics.

With a hope for a rational explanation in mind, we should not shy away from topics like these. All topics can be learned and understood given enough time and nobody should get the upper hand in a discussion just because he or she uses big but undefined words—like pointing to the dragons on the map, as if they would give them power to make proclamations about the world without having to base them on testable facts.

3.8 Chaos Theory

> Nature uses only the longest threads to weave her patterns, so each small piece of her fabric reveals the organization of the entire tapestry.

—Richard Feynman, *The Character of Physical Law*

Question

How can structure emerge from chaos? Why isn't everything just "noise"? And how does "complexity" differ from "chaos"?

Example

Imagine a droplet of honey falling into a bowl of yogurt. After you stirred the yogurt a few times, circles of long threads of honey emerge. Some more stirring and it seems like the honey drop is distributed more or less evenly in the yogurt. But the situation has not necessarily become chaotic or random: if it were possible to stir the yogurt in the opposite direction in exactly the same way, you would end up again with a single drop. And that is how nature can best be described: as a sea of waves representing an infinite series of stirrings or foldings.

To better understand the concept of chaos, let us look at a selection of examples where we intuitively expect them to be random, while the phenomena are actually based on simple repeating rules. If we discover these rules, we can explain nature's complexity without having to fall back on ideas of randomness or supernatural causes.

3.8.1 Fractals

When looking at trees, we see fine-grained structures. Like our hierarchy tree of concepts from *Philosophy for Heroes: Knowledge*, this indicates that there is a lot of information hiding within those structures. We assume that their appearance must be stored somewhere in their genetic code. But that seems unlikely given that trees branch out so much over time.

> **FRACTAL** · A *fractal* is a self-similar pattern created through repeatedly applying the same rule on itself.

A simple example of a fractal is the "Koch snowflake" where you start with a triangle and add a smaller triangle at each side. Then, you repeat that process with smaller triangles, and so on. If you repeated that a hundred times, the total circumference would be as long as the distance between the Earth and the sun. The result eerily looks like snowflakes we know (see Figures 3.18, and 3.19):

Figure 3.18: A snowflake (image source: shutterstock).

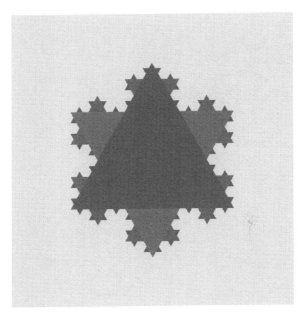

Figure 3.19: The third iteration of a fractal resembling a snowflake (image source: shutterstock).

As this is the simplest way to create scalable structures, fractals can be found in nature at many places. And having scalable rules is essential for plants which can grow only as nutrients and sunlight allow them to grow. With fractals, a tree does not have to encode branching structures in its genetic code, it simply has to make sure that branches, when reaching a certain size, branch further. Having such a pseudorandom pattern also increases the amount of light a plant can absorb by minimizing the amount of shadow the individual leaves and branches throw on other leaves. But instead of storing the whole tree structure in the genetic code, it stores only the rule for how the branching should happen and applies that rule repeatedly when growing (see Figure 3.20).

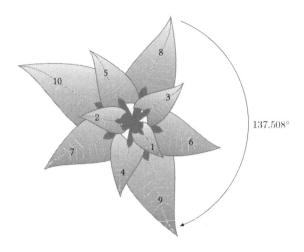

Figure 3.20: Circular distribution of plant leaves using the golden cut.

Remembering what we have talked about in the first book, *Philosophy for Heroes: Knowledge*, this also connects to the golden ratio, a numerical relationship that looks aesthetically pleasing to our eyes. This is because the underlying Fibonacci sequence $(1, 1, 2, 3, 5, 8, 13, 21, \ldots)$ of the golden cut $(\frac{5}{3}, \frac{8}{5}, \frac{13}{8}, \frac{21}{13}, \ldots)$ is also built on a repeating function, namely adding the two previous numbers $(1 + 1 = 2, 1 + 2 = 3, 2 + 3 = 5, 3 + 5 = 8, \ldots)$.

3.8.2 The Magnetic Pendulum

Imagine a pendulum and three magnets (see Figure 3.21). You release the pendulum and it swings back and forth, its path always slightly affected by the nearest magnet it swings by. Eventually, it will come to a rest above one of the magnets. You keep on repeating this experiment but no matter how often you do, you seem to never be able to predict where the pendulum will arrive at the end—some obvious starting positions like right above one magnet aside.

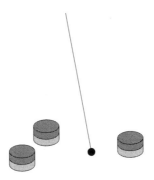

Figure 3.21: A magnetic pendulum swinging back and forth between three equally strong magnets.

If you made note of the starting positions each time you released the pendulum and assigned a color to each magnet, you would end up with a picture that looks like someone has thrown buckets of paint onto a canvas. It seems to be a product of chance, not of a deterministic cause or pattern. But examining it more closely, we do see some form of organization, which seems to be symmetric in three directions. It still could be buckets of different colors of paint having been thrown onto a canvas and then mirrored along its three (or six if you include the mirroring of the left and right side) axes.

CHANCE · The cause of an effect when no other cause could be determined.

Despite having only a simple setup of three magnets and a pendulum, the resulting image (see Figure 3.22) shows significant complexity. It is that chaotic because each swing multiplies any change of the initial starting position. The more swings the pendulum will make (the further it is away from any magnet), the harder it is to predict the final position based on a previous experiment, no matter how small the difference of the starting positions. Each swing multiplies initial inaccuracies in putting the pendulum at the same starting place as before.

Figure 3.22: Starting positions of the magnetic pendulum marked with a color corresponding to the position where the pendulum will stand still at the end of the experiment.

[Chaos is when] the present determines the future, but the approximate present does not approximately determine the future.

—Edward Lorenz

CHAOS THEORY · The *chaos theory* states that small differences in initial conditions can yield widely diverging outcomes. For example, given enough repetitions, the effect of a butterfly flapping its wings on one side of the world might cause a hurricane on the other side of the world. Beyond this butterfly effect, the chaos theory also deals with patterns that emerge from an apparently chaotic system.

3.8.3 The Raisin in the Dough

A similar example would be kneading dough. In our case, the kneading process would involve rolling out the dough, then folding it in the middle, and repeating this process. Imagine putting a raisin into the dough, and then doing the rolling out and folding process ten times. Where is the raisin?

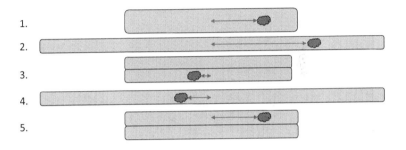

Figure 3.23: An illustration of how the parts of a dough move when it is kneaded, folded in the middle, kneaded again, and so on.

Let us say that we put a raisin at position 0.9 (0 being the left side, and 1 being the right side of the dough, see step 1 in Figure 3.23). When we roll out the dough (step 2), it is at 0.8 because the rolled out dough is twice the size of the folded one. Then, we fold it in the middle (step 3). It is now at position 0.4. Rolling it out again (step 4), it lands at 0.8. Folding it again, and we are back at 0.4. Continue this process, and the raisin would jump forth and back between 0.4 and 0.8. It has reached an "attractor" position and we can predict where it will be in the future after a certain number of repetitions.

Now, let us imagine that we made a small mistake and we started instead at 0.9001. Can we reuse the result of our previous folding or calculation to predict where this raisin will land? It will land on very similar spots for a few foldings. But after around ten foldings, the raisin will be at a very different side of the folding than it would have been if we had started at 0.9000. Ultimately, only positions that are part of an attractor (see below) can be used to predict where a raisin starting at a different position will land: if we started at 0.6, 0.4, or 0.8, we can reuse the result; if we started at 0.89999 or 0.90001, after a number of foldings, we can no longer predict where the raisin will turn up without redoing our calculation.

So, while the end result seems to be random, both the end positions of the pendulum as well as the raisin can actually be predicted. But both are very sensitive systems so that even the smallest change might have an effect on the end result. In this case, it was a simple rule (the pendulum moves according to the three magnetic forces or the folding of the dough, respectively) which, when applied multiple times (swings or foldings), produced a complex pattern.

3.8.4 Attractors

When looking at the universe not as a set of particles at certain positions, but particles that are wiggling all the time, the "fixed points" we attribute to particles need to be replaced by attractors. Something that looks like it is staying at one place is actually moving around but ultimately will return to the original setting, only to return to moving around in the same pattern. This is what we have seen with the raisin at position 0.4: The raisin will jump back and forth between 0.4 and 0.8 after each folding and rolling out. Thus, 0.4 and 0.8 are attractors. Another example of an attractor was the magnet. Obviously, the position right above a magnet is another (albeit simple) attractor. Other than the name suggests, though, the magnets in that example would *not* themselves be attractors; attractors are just the rules that are repeatedly applied with each swing.

ATTRACTOR · Repeatedly applied rules or laws can eventually loop. In chaos theory, this kind of a loop is called an "*attractor.*"

In a certain way, you could look at rivers as attractors. It rains and the empty riverbed slowly fills until it reaches an equilibrium of water flowing into the riverbed and water flowing away as part of the river. The river is the stable state, the "attractor." Likewise, rain starts out with a few droplets until it reaches a stable state of a certain amount of water falling down per minute. Similarly, you could describe the waves on the sea as a stable attractor. Sure, they move around, but the sea returns to its original state after each wave.

We see the same idea with light or sound, which we do not describe by absolute positions, but by their attractors, their wavelengths. For example, a note in music is not a singular event or particle, it is a pattern of how air pressurizes and depressurizes back and forth. We find this idea also in computer programs that loop. In old TV screens, there was a screen with a single cathode ray highlighting individual points on the screen with a certain color. This highlighting was done in lines and so fast that instead of individual lines or points lighting up, our eyes could see only a full picture.

Idea

Chaos theory explains the complexity in nature by pointing out that it is the result of repeatedly applied (simple) rules or natural laws. Stable elements within a chaotic system are called attractors. They are the result of a looping repeatedly applied rule.

At the end of the next chapter, a picture will emerge of what thoughts are. Instead of singular states of individual neurons, you can look at them as attractors—signals swinging back and forth between a number of neurons like the magnetic pendulum hovering over one magnet.

3.9 The Big Bang

Imagine an infinite sea of energy filling empty space, with waves moving around in there, occasionally coming together and producing an intense pulse. Let's say one particular pulse comes together and expands, creating our universe of space-time and matter. But there could well be other such pulses. To us, that pulse looks like a big bang; in a greater context, it's a little ripple.

—David Bohm

Why do we observe the sky? Because the universe provides us with rich stories from a time long ago which provide us with experiments, comparisons, and measurement tools—just think of stars, black holes, or quasars. Being so far away, not only do those celestial objects tell us how things are, they also tell us how things *were*. This allows physicists to wonder about the consistency of laws and constants of the universe. Just because we formulated a law does not mean that it needs to hold true forever.

For the same reasons, physicists are interested in evolution on Earth: how early lifeforms on Earth developed can give us some hints about what natural laws billions of years ago were like. But what if we go further and further back?

Figure 3.24: Illustration of the theory of the big bang (image source: shutterstock).

At the beginning, there was the "big bang" (see Figure 3.24). The theory is that at one point, a singularity exploded into particles, which over time formed into atoms, and later stars and planets. Before the so-called "big bang" that brought our universe into existence (at least that is a current scientific theory), there was "nothingness." If we start from our axiomatic system with entities having an identity in the form of properties, "nothingness" simply refers to a space where no entities exist—hence that space has no properties. While we cannot use pure philosophy to decide what exactly existed before our universe, we can make general statements about entities, properties, and their effects. By applying philosophical principles of Objectivism from *Philosophy for Heroes: Knowledge*, we can be sure that there is no beginning of the universe; even the so-called "big bang" would be merely a result of the properties of "something" which existed before the big bang and which had the properties to create a big bang. If there had been pure "nothingness" before the big bang, that nothingness would have had no properties that could have caused a big bang.

We are ultimately faced with the same issue as when we began to discuss philosophy: we first need to clearly define what we are speaking about when we use words like "big bang," "nothingness," or "universe." We have to be careful not to take on a view of the universe that is based on pure linguistics or intuitive interpretations of the words and instead take care to start from a common, clear basis of definitions.

3.9.1 Nothingness

The intuitive understanding of nothingness is that when you have a bowl of apples and empty that bowl, there is "nothing" left in the bowl. Of course, scientists discovered early on that such a bowl is actually not empty, as there is still air in the bowl. If you take any space and pump out the air, you are left with "true" nothingness: a vacuum. But this view of constructed nothingness is based on the classical view of physics. At this point, I want to stress that as students of reality, we need to get away from the idea that everything starts with our intuitive understanding of the world. We need to be careful to be objective at all times, especially when it comes to non-intuitive questions from philosophy and physics. When one removes all entities from a box by pumping out the air and creating a vacuum, that does not mean that the space inside the box is left with no properties. While you might be unable to "move" space in the conventional sense of a thing, it would still fit our definition of an entity. There is no requirement for the universe having to have a clean, property-less canvas on which it draws its entities; the canvas itself can have properties.

> In quantum gravity, universes can, and indeed always will, spontaneously appear from nothing. Such universes need not be empty, but can have matter and radiation in them, as long as the total energy, including the negative energy associated with gravity, is zero.
>
> —Lawrence Krauss, *A Universe From Nothing*

One of these properties of space is that it can spontaneously create two particles that cancel each other out energetically. This has been shown to happen in a number of experiments.

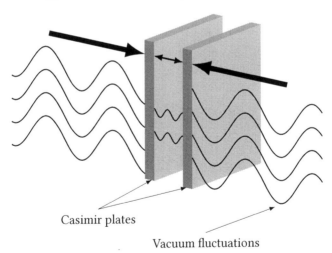

Casimir plates

Vacuum fluctuations

Figure 3.25: Illustration of Casimir forces on parallel plates that occur in a vacuum.

Example

You can create a vacuum and place two metal plates, facing each other, in the vacuum. According to classical physics, nothing special should happen. Measurements have shown, though, that there is a force to push those plates away from each other (or pull them together, depending on the setup), despite being in a vacuum and despite no other forces being at work (Casimir effect, see Figure 3.25). A macroscopic model that shows the same effect is a vibrating water bath into which you put two plates: the waves reflecting from the sides of the water bath push the plates together.

Example

A black hole is basically a very heavy star, so heavy not even light escapes. In classical physics, anything within the so-called "event horizon," the distance at which light can still escape the gravity of the black hole, is thought to vanish in the black hole forever. But this is not what we actually observe. In fact, when particle pairs are generated and split on the event horizon with the anti-matter particle falling into the black hole while the matter particle escapes in a stream of particles, the black hole slowly loses mass: anti-matter and matter particles cancel each other out. According to the theory, what is eventually left of a black hole is a normal star. This phenomena is called "Hawking radiation."

3.9.2 The Universe

While we have argued that a part of space can be an entity because it has properties, we need to examine this more closely. What exactly is the universe? Is it everything that exists? Is it the canvas on which other entities are "painted"?

This question looks difficult to answer and depends on the context. The concept of "universe" is used in various ways. In the classical sense, the universe is everything that came into existence resulting from the big bang. That is then simply a set of entities, not an entity itself.

An alternate view is that the universe is everything that exists, but not as a set, rather as a whole entity consisting of loosely connected particles. Likewise, if we use the idea from above that the universe is a canvas, it would be infinitely large. Based on the ideas of David Bohm, studies point in the direction of infinite age of the universe[11] and the spontaneous creation of the universe from "nothingness."[12] This not only solves the problem why "our" universe is exactly how it is, but also would solve any issues arising from the idea of a big bang as a singularity and a beginning of everything.

An infinitely large (and infinitely old) universe would at least explain some of the mysteries about the universe, namely why we have certain natural laws that just happen to allow for stars, planets, and ultimately intelligent life to evolve. There might be other parts on this canvas where these laws are different and where no intelligent life is possible. Maybe most of this canvas is inhospitable for life, except for a certain corner with very specific natural laws—which just happens to be our part of the universe. So, even within an infinite universe, there could be finite spaces where intelligent life can

---

[11] Ali and Das, *Cosmology from quantum potential.*
[12] He, Gao, and Cai, *Spontaneous creation of the universe from nothing.*

ponder the question why the universe is "designed" exactly in such a way that it allows for them to exist. Some of the mysteries of the universe could then be simply solved by saying that things are as they are, because if they were not, we would not be here to observe them.

> **ANTHROPIC PRINCIPLE** · The *anthropic principle* is the consideration of how the environment and natural laws just happen to support human life: only the inhabitants of those worlds that can sustain intelligent life can wonder why their own world happens to support intelligent life. If the conditions for intelligence were not met, there would be nobody wondering about it.

Did you know?

Looking into the future, the anthropic principle does not look that simple. Just because Earth made it possible for us to develop up to the point where we discovered the anthropic principle and Earth's role in the universe, that does not mean it will automatically sustain human life in the future. That is up to us. Just like we have to learn how our bodies and minds work, we have to learn how the planet works.

$\longrightarrow$ Read more in *Philosophy for Heroes: Epos*

Even though the big bang was caused by a very improbable event (a whole set of quantum fluctuations converging at one point), in an infinite universe, even very improbable events can happen infinite times. And an infinite number of big bangs would mean an infinite number of worlds where infinite copies of us sit an infinite amount of times thinking about this very question... ultimately, any idea of "identity" would lose its meaning: something without limits cannot be defined and thus has no properties. At the same time, infinity in this regard is a mathematical concept: a measurement, not a concept. Either way, at least for now, the universe still offers a lot of room for exploration.

If we ignore the larger canvas and quantum theory for a moment and focus on the universe as simply the product of the big bang, we can at least make a statement: *this* universe is finite. It is as large as the extent to which particles traveled since the big bang. If there were no big bang and if the universe were infinite, the night sky would be either brightly lit with the light of "infinite suns" or the suns would all have to be so far from each other that, from an observer's point of view, there are only a limited number of suns visible at any point in time (which does not correlate with our observations).

Yet, there could be an alternative. The hypothesis above assumed that in the infinite universe, the laws of the universe differ from place to place. There could be an infinite space where life, chemistry, maybe even basic physics are not possible. For example, if the gravitation constant were different, hydrogen atoms would not attract to each other and no stars could form. Likewise, there are dozens of other constants in the universe that are set just "right" to allow life. Some might assume that this is due to a supernatural creation, others assume that this is simply so because we are in a part of the universe where these constants happen to be that way. Like ecological niches where certain forms of life thrive, you could imagine the universe like an infinite jungle with different constants providing home for stars and even biological life. At this point, it is up to future philosophers, cosmologists, and physicists to expand our conceptual understanding of our existence.

Idea

What we call "nothingness" is still a region of space where the laws of physics apply. And in physics, nothingness is not stable. New particles are generated and destroyed at all times. It is conceivable that our "big bang" happened by accident some place within an infinite universe.

The book could end here if it were not for a strange oddity. Out of this jiggling chaos of the universe, life emerged on the third rock of a solar system located on a minor spiral arm of the Milky Way (see Figure 3.26)—a galaxy consisting of around 200 billion stars—which itself is part of a group of 100 other galaxies making up the Virgo Supercluster, which itself is part of the Laniakea Supercluster, consisting of 100,000 other galaxies, which again is part of the observable universe—which in turn spans more than 90 billion light years.

Let us now narrow our focus on the part of the universe that does support life and let us look at the bridge between physics and biology. How could life emerge from non-living matter?

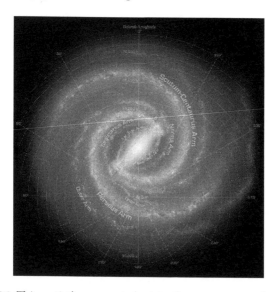

Figure 3.26: This artist's concept depicts the most up-to-date information about the shape of our own Milky Way galaxy. We live around a star, our sun, located about two-thirds of the way out from the center. Credits: NASA/JPL-Caltech/R. Hurt (SSC/Caltech)

Chapter 4

Evolution

 It makes one wonder what the evolutionary tree of this idea [the theory of evolution] would look like, were it an organism that could be mapped out by fossil record rather than words. The concept is one that faded nearly into obscurity, only now to be revived with slight mutation. What I personally gather from this is that survival of ideas depends less on the actual quality of the idea, but rather the climate into which it is introduced. Quite literally, survival of the fittest, but not necessarily the best.
—*Aquinas and Evolution*

Figure 4.1: Three photographs of Charles Darwin, father of the *Theory of Evolution*, at three different times of his life (image source: shutterstock).

In the previous chapter, we learned about the inner workings of the universe. True, there is still mystery out there. But if we accept the results from physics and chemistry as given, what can we learn about the origin of life, evolution, and ultimately our own creativity? The discussion will prepare the way for the next book, *Philosophy for Heroes: Act*, where we will examine the inner workings of our psychology and our values.

Evolution itself still evolves by adding more abstraction layers to it (DNA, multicellularity, sexual reproduction), by creating conditions for faster evolutionary processes within an organism (immune systems, the mind), and finally by having such an organism recreate new artificial organisms based on these principles (artificial life). Each of those *evolutions of evolution* leads to more complex organisms on Earth.

Significant steps along this path include:

- **3.8 billion years ago**: Earth cools down, meteoroid bombardment ends, first protocells with RNA appear, deep sea hydrothermal vents provide basic metabolism for protocells.

- **2.45 billion years ago**: Cells produce oxygen toxic to most other existing bacteria, massive climate shift, photosynthesis begins (Great Oxidation Event).

- **2 billion years ago**: Specialized cells appear, genotype / phenotype differentiation begins, cell machinery separates into DNA, RNA, and proteins.

- **1.5 billion years ago**: Multicellular organisms appear.

- **1.2 billion years ago**: Sexual reproduction and crossing over of DNA speed up evolution.

- **500 million years ago**: Adaptive immune systems based on an evolutionary algorithm appear, first animals appear.

- **220 million years ago**: Mammals appear.

- **100 million years ago**: Primates appear.

- **65 million years ago**: Asteroid collides with Earth (Cretaceous–Paleogene extinction event), climate shifts, dinosaurs become mostly extinct, rise of the mammals begins.

- **8 million years ago**: Last common ancestor of humans, chimpanzees, and gorillas disappears.

- **2.5 million years ago**: Humans appear.

- **300,000 years ago**: Modern humans with a "Darwin machine" (a high-speed evolutionary algorithm) in the brain appear.

- **70,000 years ago**: Writing is invented.

- **2,000 years ago**: First mechanical computer for astronomy (Antikythera mechanism) is developed.

- **1,000 years ago**: Scientific method is developed.

- **1543**: Scientific Revolution begins.

- **1859**: Theory of evolution is published.

- **~100 years ago**: Theory of relativity and quantum theory are published.

- **1950s**: Theory of DNA is published, digital computer is developed, cellular automata, and evolutionary algorithms are used for the first time on the computer.

- **1960s**: First human lands on the Moon, development of the Internet starts.

- **1970s**: Theory of adaptive immune systems and theory of memes are published.

- **1980s**: Theory of neural Darwinism is published.

- **1990s**: Artificial life is created on the computer, World Wide Web is launched.

Coming back from this bird's eye perspective to the ground, when people think about the theory of evolution, it is typically Charles Darwin who comes to mind—he is celebrated as the one who came up with a *revolutionary* idea. But people tend to remember only the first (or last) participants in a long series of events. For example, take the first humans on the moon: Neil Armstrong and Buzz Aldrin are well known, but who remembers the person staying in orbit, Michael Collins?

Michael Collins was the third person on the Apollo 11 mission, remaining in orbit while Armstrong and Aldrin descended onto the moon. The mission would have been impossible without him, just as the mission would have been impossible without the people building the moon lander, the computers, and the rocket, or the people managing the operation on the ground. "Heroes" would not have been able to land on the moon and successfully return. The operation was too complex to be achieved with a singular heroic effort; thousands of things had to be just right. Sure, the astronauts were risking their lives, but so did the people on the ground, every day they drove to work.

Collins feared that something would happen and that he would be the only one to return, with all the spotlight on him, then being "a marked man." In a recent interview, at the age 78, he said that he is bothered by today's inflation of heroism and adulation of celebrities, adding that he is no hero and that "heroes abound, but don't count astronauts among them. We worked very hard, we did our jobs to near perfection, but that is what we had been hired to do."

Learning about the complexity of the moon landing relativizes the role of the astronauts, just like learning about the centuries-long evolution of the theory that Darwin made famous relativizes his role. Neither the moon landing nor the theory of evolution is the result of magic or a singular heroic effort. Each can be understood by looking at the whole chain of industry, of scientists, and of ideas.

We are often taught only about the last element of a long chain of events. And those not familiar with a subject see this last element as almost supernatural. In science, it is easy to overlook how theories were developed over centuries. This applies to the theory of evolution, which is difficult to understand if one simply jumps over thousands of years of scientific progress and focuses only on the final result.

In ancient times, during Homer's era (ca. 750 BC), life was understood as the result of the action of whimsical, inconsistent gods. In this world, the rather primitive statement by Thales, "The first principle and basic nature of all things is water," launched a dramatic shift in people's minds. No longer were people discussing the moods and personalities of human-like gods. Instead, these early philosophers looked for patterns in nature to explain natural events. Rather than relying on stories and myths, people were beginning to test truths on a first-hand basis.

One of the first documented thinkers who promoted this idea—that the phenomena of nature cannot be explained by supernatural gods or magic, but by observable facts—was Anaximander (610 - 547 BC). Starting with Thales' idea that water is the origin of all things, and the observation of humidification and cloud formation, Anaximander concluded that in earlier times, the Earth must have been covered by water. The existence of fossils further strengthened his view, which led to the conclusion that humans also had to have emerged from water.

Not long after, Empedocles (490 - 430 BC) offered an explanation of why organisms in nature look as if someone had designed them for a specific purpose. Those organisms happened to have properties that allowed them to survive in their environment. Those that did not died and hence, were not part of nature anymore. He also held the view that life could have developed without an underlying purpose or a godly creator.

Further support of this idea was provided by Aristotle (384 - 322 BC). Like Charles Darwin 2,000 years later, he was an explorer of nature, analyzing and classifying more than 500 different animal species. He recognized that animals' properties were specifically adapted to their environment. He disagreed with Empedocles, though, as to whether or not they had a higher purpose.

Later, in the Roman empire, Lucretius (97 - 55 BC) took interest in the subject and based his work *De Rerum Natura* [1] on the writings of Epicurus. The book describes the universe as a purely mechanistic entity, without supernatural influence. This idea thrived within

---

[1] "On the Nature of Things," published by Cicero [106 - 43 BC] after Lucretius' death.

an environment of the then-popular Stoicism, the view that to attain happiness, you need to understand nature. Contrary to the theory of evolution, though, people still believed that everything happened for a reason, that the world was designed for a purpose (teleology).

With the deterioration of the Roman Empire came the attempt to keep the empire together by raising Christianity to the state religion. Augustin of Hippo (354 - 430)—one of the so-called "Church Fathers" with a strong influence on the philosophy and theology of Christianity—argued against the idea of literal interpretation of the Bible, claiming that new species can develop.

The Roman Empire fell in 476, and it took until the 9th century for learning centers in the Middle East to be rediscovered and to translate old Greek and Latin books. Similarly to Aristoteles, Al-Biruni (776 - 868) categorized in his *Book of Animals* over 350 different animal species, their environments, and their places on the food chain. One of his notable discoveries was that animals are constantly in a fight for survival, and that successful properties of the animals are inherited to the next generation, resulting in adaptations and even new species.

The zenith of Arabic scholarship was with Nasīr al-Dīn Tūsī, a Persian polymath (1201 - 1274). He stated that "The organisms that can gain the new features faster are more variable. As a result, they gain advantages over other creatures. [...] The bodies are changing as a result of the internal and external interactions." He even theorized that humans are but a middle step of an evolutionary stairway, with animals as the precursors, and humans with spiritual perfection as the successors.

As climatic conditions in Europe improved during the High Middle Ages, scientific progress returned. From the Middle East, translations from Arabic books, old Greek and Latin writings, and even scientists returned to Europe, creating the foundation for people like Thomas Aquinas (1225 - 1274). His view was that God provided an objective world with cause and event in an endless loop, and that animals had a god-created potential to develop into new species. Besides the initial potential of nature to create this diversity, there would be no further godly interventions in this progress.

The Crusades and the Mongol Invasion in the Middle East, and the subsequent destruction of libraries and the fabric of society, left Europe as the keeper of knowledge. With the beginning of the Little Ice Age (1300 - 1750), a strengthening of the Catholic Church, the persecution of heresy by religious inquisition, as well as with a lack of literacy in the general population, it took until René Descartes (1596 - 1650) for a revival of a mechanical view of the universe, as opposed to a supernatural one.

Benoît de Maillet (1656 - 1738) was a student of geology. Starting from a theory by Descartes, namely that the Earth had originally been entirely covered by water, and by studying fossilized shells embedded in sedimentary rocks on mountains high above seal level, he concluded that the Earth must have been created not by a singular act, but by a slow, natural process. He estimated the true age of the Earth at around 2 billion years and assumed that humans must have been developed from animals that came out of the water.

- Anaximander provided the idea that natural phenomena can be discovered through observation instead of relying on the idea of godly intervention.

- Milet theorized that life emerged and originated from the sea.

- Empedocles had the idea that the life forms weren't just there, but that they had their properties in order to survive in their environment—without any higher purpose.

- Aristotle conducted plant and animal gathering and classification work, agreed with Empedokles about their adaption to the environment, but assumed a higher purpose.

- Lucretius and Cicero held the views of a mechanistic universe.

- Augustinus had the idea that new species could evolve.

- Al-Biruni again, like Aristotle, was someone who categorized animals (which seemed to be the key in understanding evolution) and proposed the idea that animals are in a fight for survival and new traits could be inherited to a new generation and that, in this way, new species could evolve.

- Aquinas held the view that life forms have an inherent potential to develop into new species—without any intervention by God.

- René Descartes again picked up the view of a mechanistic universe.

- Benoît de Maillet studied geology and found the Earth not static and original as it was created by God, but in a constant change. He estimated the age of the earth at 2 billion years and deduced that animals evolved from life forms from the sea.

Finally, in the following two centuries, the idea of evolution took off.

- First, while fossils were found and considered the remains of life forms from ancient times, the fossil record was very sparse. This gap began to close thanks to industrialization, which increased the need for professional geologists, and coal mining uncovered new fossils.

- Second, the comparison of specimens from different geographical areas provided evidence for a relationship between the species: the greater the distance between two species, the more they differed from each other.

- Third, the ancient writings became known and available to more and more people. Book printing grew from a few hundred titles per year in the 17th century to thousands of titles per year in the 18th century.

- Fourth, the general science of categorizing the natural world created the foundation for further scientific theories including the theory of evolution by filling gaps, focusing the research on the questions that remained open.

THEORY OF EVOLUTION · The *theory of evolution* states that the process of evolution tends to create systems in each new generation that are better adapted to the environment (same or higher rate of procreation compared to the parent generation).

With this background in mind, Charles Darwin's (1809 - 1882) *Theory of Evolution* in his book *Origin of Species* looks like a much smaller step than when taken on its own. It did not come out of the void and we have to remember those who paved the way. Charles Darwin's research certainly was a heroic act, given the resistance he faced (and his work still faces). His integration of all the pieces that were available to him, including his own research that he conducted when taking part in the voyages of the HMS Beagle around the world, was revolutionary—despite having an *evolutionary* record.

Likewise, the history of the theory of evolution did not end with Darwin. There were (and still are) many gaps being filled. Advances in other fields of technology opened the door for genetic research and we are only now slowly beginning to understand the code in which life is written.

Science is a collaborative enterprise spanning the generations. We remember those who prepared the way, seeing for them also.

—Carl Sagan, *Cosmos: Blues for a Red Planet*

4.1 Basics of Evolution

> **Question**
>
> What are the three major properties a system needs to have in order to evolve?

> Still, it needs to be said that the light of evolution is just that—a means of seeing better. It is not a description of all things human, nor is it a clear prediction of what will happen next.

—Melvin Konner, *The Tangled Wing*

The purpose of the theory of evolution is to explain the wide array of life forms on Earth. Knowledge about the origin of species allows us to better understand why a particular organism has certain properties. While this is often obvious by looking at the environment of an organism, for example by looking at the teeth and the food, or the leaves and the precipitation and angle of sun rays, more complex behavior seems baffling, especially that of humans. Why and how did humans develop? Why do we have such a big brain? What is the reason for our behavior? All these subjects can be better understood when we understand our past—how life, and humans in particular, developed on Earth.

> He who does not understand the uniqueness of individuals is unable to understand the working of natural selection.

—Ernst Mayr, *The Growth of Biological Thought*

Despite its very basic mechanics, evolution is a complex concept, requiring that we approach it in several steps. We will start where the first book left off: entities, properties, systems, aggregates, etc. This provides the philosophical foundation of the theory of evolution. In later sections of this chapter, we will take a look at chemical evolution, artificial life created by computer and ultimately the origin of life on Earth, evolution in our immune system, and evolution in our head.

4.1.1 Selection of Systems

Before we discuss evolution in detail, it is important to note that parts of the evolutionary process can be applied not just to living organisms, but also to *any* system—for example to chemical systems, simulated life forms in a computer simulation, or thought patterns in the brain. One of these parts is the selection process. To understand selection, let us revisit some of the definitions from the first book and add some new ones:

AGGREGATE · An *aggregate* is a number of entities that have a reciprocal effect on one another, so that they can be considered collectively as their own entity (e.g., a cup full of water—all water molecules interact with each other).

STRUCTURE · A *structure* is a description of required properties, dependencies, and arrangement of a number of entities (e.g., cube-shaped).

SYSTEM · A *system* is an aggregate with a definite structure (e.g., an ice cube is a system of frozen water molecules).

PROCESS · A *process* describes the mechanism of a cause working to an effect (e.g., if you put an ice cube into a glass of water, the cooling of the water is the process).

TIME · *Time* is a measurement tool to put the speed of processes in relation to each other.

PROCREATION · *Procreation* is a process by which a system creates (on its own or with the help of the environment) a new entity with a similar or (preferably) the same structure as itself.

GENERATION · A *generation* is a set of systems during one cycle of procreation.

GENOTYPE · The *genotype* is a system that is the blueprint for the phenotype.

PHENOTYPE · The *phenotype* is the actual body of a life form. Changes in the phenotype generally do not have effects on the genotype. Generally only the phenotype interacts directly with the environment.

MUTATION · A *mutation* is a change of the genotype of a life form. This change can, but does not necessarily, have consequences for the phenotype.

4.1.2 Mutations

Mutations, as our common knowledge suggests, are generally bad. No matter their effect for the overall population, for the individual life form, mutations are unwanted accidents. They happen, but they are not intended. An individual life form which is adapted to its environment does not want change. If the parent generation already has well adapted genes, the last thing they want is their offspring to have random changes. The only situation where (more) mutations are welcomed is when the life form is in a tight evolutionary race like a virus with its host—we will talk in Section 4.5, "The Red Queen Hypothesis," about this situation and how life copes with it. But usually, the environment changes slowly enough that mutation is not a limiting factor.

Sometimes, multiple repair and protection mechanisms are bypassed by a mutation while it also is mutated in a way that it does not stop reproducing: that is cancer. To understand cancer, you have to understand multicellular organisms. For a single-celled life form, there is no such thing as cancer. Cancer is simply an unlimited self-replication of a cell. While this is exactly what you want for a single-celled life form, it possibly destroys the structure within a multicellular organism. If all of our cells divided uncontrollably, we would not survive. Our life depends on a highly organized and controlled process of cellular growth and death. Cancer for us occurs when those processes fail for a single cell.

> **Example**
>
> When making a copy of themselves, most of our cells lose a few genetic base pairs at their ends. Once they have lost a certain amount of code, they stop duplicating themselves. Any damage of a defective cell would then be limited by its remaining duplication cycles. Any form of cancer cells would need to have undergone a mutation that fixes the ends of its genetic code after each duplication (Shay and Wright, *Role of telomeres and telomerase in cancer*). In a similar fashion, there are other protection mechanisms in the cell that a cancer cell has to overcome. For example, scientists have found that there is a single gene regulating multi-cellularity in organisms. It acts like a brake to stop multiplication at the right times. If that gene is defective in a cell, it might end up multiplying indefinitely and cause cancer (Hanschen et al., *The Gonium pectorale genome demonstrates co-option of cell cycle regulation during the evolution of multicellularity*).

4.1.3 Selection

> **FITNESS LANDSCAPE** · The *fitness landscape* is the sum of all environmental influences on an entity. For example, if you are sifting sand, a riddle screen lets small particles of sand fall through while larger stones are retained. In this case, the riddle screen and the shaking of the riddle screen would be the "fitness landscape." In nature, the fitness landscape would simply be the environment over time, including all other life forms, the climate, etc.

> **SELECTION (EVOLUTION)** · *Selection* is a process where some (or all) of a set of systems of similar structure are retained while the rest are discarded or destroyed. Which ones are retained and which ones are discarded depends on the relationship between the structure of the individual systems and the fitness landscape. In the example of the riddle screen, the screen lets small sand fall through while it "selects" larger stones.

Selection is a central element of evolution. It connects systems with environmental factors. When the environment changes, different systems are selected. In order for selection to work—or to make any difference or sense—there needs to be variation. As such, the mutation rate of a population is caught between two conflicting forces: the tendency to replicate what works and fix any copying errors, and the need for some variety in the population in order for a population to evolve.

Nature's solution for the conflict between variety and preventing errors was to keep mutations as low as possible, while using sexual recombination at the same time. Like mutations, sexual recombination keeps up the variety within a population. But it also kept individual genes intact, reducing the probability for destructive changes: it simply recombined working copies, resulting in a varied, but very likely functioning organism. It is hard to compare it to any everyday concept because it is a very unique solution. The closest analogy might be editing a movie where you cut and rearrange parts instead of trying to change individual pixels.

Idea

A population needs to have variations in order to evolve.

Example

Imagine trying to sift identical stones with a fine riddle screen: nothing goes through. Or imagine trying to sift sand with a riddle screen: everything goes through. Likewise in nature, without change in the population, a population cannot move in any direction. If you use a finer riddle screen, more sand is retained. This changes the ratios between the different types of systems in the environment (for example the ratio between large stones and fine sand). Such changes could be caused by either selecting more (and discarding fewer), selecting fewer (and discarding more), or by selecting other (and discarding other) systems.

Example

Imagine stones in a stream. The stones with the highest flow resistance experience the highest degree of erosion. They are "discarded" while rounder stones with lower flow resistance remain the same. Obviously, the "discarded" stones are not actually discarded but transformed into stones with lower flow resistance simply because erosion slowly removes the edges of those stones until they, too, become round.

The significant point is that selection applies to all things in the universe. It is simply a process describing the change of systems over time due to external influences by other entities. A system with lower "durability" will by definition not endure as long as a system with higher "durability," so after a while, you will see the ratio of systems with higher durability increase. For this principle, we do not even have to look much at nature, as it is very much a logical process based on the properties of the involved entities.

But for evolution to happen, another component is necessary. The system needs to be able to "clone" itself and replace the systems that were discarded during the selection process. Looking again at our example with the stones in the stream, you can imagine that over time, the stones get smaller and smaller until only sand remains that gets easily pushed down the stream. With no new stones being added to the stream, all stones eventually will have turned into sand and washed away—even the ones that were least resistant to the flow. To clarify, sand and stones in a riddle screen do not "procreate" in any form, so while they undergo a process of selection, they do not undergo a process of *evolution*. You could argue that the stones in a stream adapt to their environment by replacing themselves with a rounder version, but this is a borderline case as erosion is very much a one-way street.

Idea

In order for evolution to occur, the life form has to be able to increase the number of copies of itself in its environment.

4.1.4 Selfish Genes

As stated previously, for evolution to work, selection needs to be—as the term suggests—selective. This means that selection needs to be based on an individual life form's properties and not on a whole population of life forms. It is not enough that a change improves the ability for a life form to survive. It needs to improve the life form's likelihood of survival *relative to other life forms.*

For example, if a life form has a mutation that produces chemicals that speed up the process of procreation but distributes these chemicals among all other life forms, every life form nearby profits, no matter whether they had the beneficial mutation or not. This is similar to the situation in an economy. Money, status, or reputation is given preferably to those people and companies that have methods more adapted to the environment (the market). A well-run restaurant with better service will attract more customers than a poorly run restaurant. If, at the end of the day, the better restaurant had to share its earnings with all other restaurants, it might have a harder time "evolving" and expanding its business.

> **Example**
>
> An exception to the rule of having to work for one's advantage are beehives. They consist of infertile worker bees and a single fertile queen. In this case, in terms of evolution, they can be seen as one large organism where individual worker bees have no problem sacrificing themselves for the greater good of the hive—if its queen survives, the worker bee's genetic material survives. This idea is best explained in Richard Dawkins' *The Selfish Gene* where he stresses how organisms tend to act primarily in favor of the survival of their genes, not necessarily of their own bodies.

With that, we arrive at a basic definition of evolution:

> **EVOLUTION** · *Evolution* is the combination of the process of selection together with a system of cloning or procreation.

Idea

In order for evolution to occur, the life form has to be able to increase the number of copies of itself in its environment. In addition, the resulting population of life forms needs to have variation, and each life form has to be able to operate for its own advantage.

4.1.5 Randomness

There are many slopes leading up a mountain, but there are but a few mountain tops.

Question

If it is all about variety, why do we still intuitively think that mutations are the driver of change and evolution?

It is easy to mistake mutations for somehow being the drivers of evolution. I cannot blame people who might do that. Sure, this misrepresents the theory of evolution, but it is ultimately the job of the proponents of the theory of evolution to make the right points. Similarly, it is easy to make the mistake of explaining evolution as if one type of organism mutated into another. But that explanation is like the *many worlds interpretation*, where our universe's past has but a single deterministic history, but at each step, it looks like it could branch off into an infinite number of possibilities. Likewise, when looking back at your own ancestry, you see a long, unbroken string of ancestors who all successfully survived and procreated.

So, how do we approach this issue? Yes, mutation is part of the evolutionary process. And if we look back, we see ourselves as the product of a string of mutations. The problem with this approach is that it is a very anthropocentric view. What we need is a different perspective; we need to look at it not from the position of the "survivor," but that of an outside observer.

From the outside, we see a population of individual organisms, with mutations causing diversity within the population. Over generations, the gene pool of the population slowly moves in a direction in which the organisms are better adapted to their environment. From this perspective, the driving force clearly is the environment. For example, let us say that within a population, there is one individual with a resistance to a certain disease. Over generations, his descendants will be more successful than other members of the population, so the gene with the resistance slowly spreads within the population until the majority has it. From there, another mutation might occur and spread in the same manner. Looking back, it seems that a single line of ancestors did all the "work" to find the advantageous mutation.

Evolution can be *very* predictable. For example, in nature, the problem of how to design a body or wings with low air resistance is solved again and again similarly because the underlying physics are always the same. There might even be only one (or very few) possible optimal solutions or paths to a problem so that you can even see identical genes in non-related species.

Example

Imagine you have a mailing list with 10,000 addresses. You create two groups, sending half of them a message that you have a special ability to predict next week's stock market and that a certain share will go up. Likewise, you send the other half a similar message, but tell them the share will go down. After a week, the stock will go up or down and either way, you discard half your list. To the remaining 5,000—the people whom you told correctly whether the share would go up or down—you will send another message, reminding them about your correct prediction. And you divide them again in two groups, telling one that a certain share will go up, and the other that it will go down. After a few iterations, the remaining group of maybe 100 to 500 people will think you are some kind of genius always making the right prediction. Similarly, when looking only at the results in nature, we might think that some kind of a genius has designed life on Earth. But it only looks this way because we ignore the number of times nature has produced results that did not survive to procreate.

Another factor in the preference for believing in mutations as opposed to adaption to the environment is that it leaves the door open for random events. But even if it is agreed that the environment is the driving factor of evolution, someone could argue that changes in the *environment* are random. But as we have discussed in the first book, *Philosophy for Heroes: Knowledge*: if (supposedly) "random" events were frequent enough, we always have found a pattern in the form of a natural law. In terms of evolution, one could argue that, yes, "random" events drive evolution through mutation, but these events are not frequent enough (yet) to be discovered by science and made into a natural law. Likewise, a changing climate because of a volcanic eruption is a "random event"—but only "random" insofar as we do not have the necessary data and do not yet understand the processes that led to such an eruption.

> **Idea**
>
> It appears that mutations are the drivers of change and evolution, because we are looking at our evolutionary history as survivors. Only if we ignore all the attempts of nature that did not lead to a successful organism, life looks as if it is driven by nothing but a series of miracles.

4.1.6 Evolution as Waves

Imagine sending 100 rats into a maze and then wondering if the one that finds the way out is some kind of maze-genius. But now (after all rats have left the maze) fill it up with water, and then cause some waves at the entrance of the maze until the waves have traversed through the whole maze to the exit. Would you think that those water waves are some kind of intelligent water molecules that found the right way through the maze? Obviously not, water waves simply expand into all directions and create the illusion that they are somehow intelligent. Likewise, populations in nature should be looked at as waves, with different members of the population being at different positions of the wave. Through their genetic diversity, they also expand like a wave, checking every direction for the "exit" —a genetic code that is better adapted to their environment.

This can be seen for example in a study made with E. Coli bacteria.[2] The bacteria were spread on the left and right side of a large, rectangle shape petri dish. The petri dish was divided into different sections with no antibiotics, some antibiotics, and toward the middle 10x, 100x, and 1000x the amount of antibiotics. Over the course of 11 days, the bacteria evolved to gain more and more resistance until they hit the center. Mutated bacteria with significantly higher

---

[2]See https://hms.harvard.edu/news/bugs-screen.

antibiotic resistance are able to enter a new sector. They are then the origin of a new wave of copies that are able to spread throughout the sector until hitting the next higher antibiotics barrier.

So, again, mutations (or recombinations and any process which affects the genotype or the process of phenotypical development) only create a variety of solutions, which then are selected by the environment. If mutations were the driving factor and not just the provider of variety, evolutionary development would work without selection by the environment: life forms would develop independent of their surrounding nature. If from this random evolution life forms developed that were as adapted as we see in nature, this, in fact, would be like magic and the idea of a supernatural, driving force behind evolution would sound more rational.

The only time when mutations actually drive evolution is when the environment changes in a way that a certain trait is no longer needed by the life form. If it no longer matters if a life form does or does not have a certain gene, any mutated version of the gene has the same probability to be selected or not. Over time, more and more mutations accumulate until the original gene has become more and more rare and ultimately disappears. An example of a purely mutation-driven evolution is an *atavism*. For example, sometimes, children are born with tails because the combination of the parents' genetic remnants of a tail lead to a half-working copy of the genetic information of this limb of our distant ancestors.

4.1.7 "Macro-evolution"

> **Question**
>
> Are larger mutations ("macro-evolution") that suddenly give an organism a significant advantage an argument against the idea that variety is the driver of evolution?

While there is certainly enough literature on that idea in the *fiction* department, there is little evidence for the idea of "macro-evolution" in the real world but a whole number of arguments against it. First, there is no need for macro-mutations: all our organs can be explained by small evolutionary steps instead of large leaps. Second, any larger change to the organism affects all the parts of the organism which then have to be optimized again.

Imagine replacing the engines of an airplane with new, much larger ones. Sure, they are "better," but now the wings have to be changed and the whole model of the airplane has to be redesigned, too. Not to mention the changed fuel consumption, or the change of the maintenance procedures. That "better" plane would probably not take off from the ground. Sure, after some iterations of optimization it might, but nature does not have the luxury to wait for a few generations until an organism is "tested" in the environment. Every single iteration has to be good, as evolution cannot back-track.

Turning the question about macro-mutations around, how did we acquire abilities that were only possible with multiple changes to the whole organism? For example, for spoken language, we need a specialized brain, a specialized throat and mouth, the ability to control our breathing, and language itself. If just one thing were missing, any of these changes would be worthless. Or would they?

One argument *against* evolution is that for such (and many other) functions, we needed large jumps to accomplish them. Basically, they look as if a designer knew what he or she was doing and created individual parts that fit perfectly together. With this "designer," organs could evolve in anticipation of a newly desired trait in a later generation.

To argue against this point, it is necessary to understand how new traits can be added to an organism. The process of evolution at no point has the capability to "prepare" anything for some later generation, it needs to work now. In reality, it works the opposite way: evolution moves toward the point normally until all the conditions are right, and then a number of mutations combine the different existing abilities into a new one. Each step on the way contributes to a better organism, and each new generation afterward optimizes the newly found ability.

The point is that traits can develop even when there is neither a supernatural driver "directing" evolution, nor a specific requirement by the environment that puts a population under evolutionary pressure (like a predator). For example, spoken language certainly helped our ancestors, but it only started to actually develop once all the requirements were available. In biology, this is also called *exaptation*.

> **EXAPTATION** · *Exaptation* is the use of a certain trait for a problem or environment other than what it was originally "intended" for. An example would be feathers that started out as heat insulation, and only later were used to improve jumping, and finally for flight.

In technology, this can also be seen with 3D graphic cards: 20 years ago, when the first 3D accelerator cards came to market, nobody thought about using them for complex problems of astrophysics, biochemistry (California, *Open-source software for volunteer computing and grid computing*), cryptography, or the "mining" of virtual currencies. These functions piggybacked on the success of 3D gaming until the cards were so advanced that they could be used for other applications.

When examining the details, life looks anything but "designed." There are too many strange design decisions, for example that the nerves of our eyes sit above our retina instead below it. This leads to us having a blind spot where the nerves lead back toward our brain. A designer would have done it the other way around, like it is in octopus' eyes.

Likewise, nobody started out designing a 3D graphics card with the purpose of mining virtual currencies. Twenty years ago, that idea alone seemed ludicrous on multiple levels. But looking only at the present, the different technologies fit very well together, as if someone had planned out the whole thing from the beginning. In reality, the "planning," for example in product development, usually leads to a product that looks very much "unplanned" and clunky. Hence, there is currently a transition going on away from classical planning from start to finish, and toward a more "agile" management method with many iterations. The reason this can be better is that during the development process, a company learns a lot of new things about a product, and the market and general environment change, too.

Looking at it from a different perspective, if we gave a truly good designer the task of creating plants that would still exist in 500 million years, the designer would rely on the theory of evolution. The designer would have no way to predict how the world would change over the course of 500 million years, so any form of "planning" would be fruitless. Instead of actually designing the biology of the plant, the only chance is to give it the ability to adapt to its environment.

> **Idea**
>
> Mutations that led to larger evolutionary jumps were always the result of many smaller changes. For example, the *most recent* mutation that enabled human language was prepared for by many other genetic changes (conscious breathing, descended larynx, etc.) with their own individual advantages.

In that regard, it is important to note that life on Earth is far from stable. We know of several extinction events in the last 500 million years that caused massive reductions (50 - 95%) in biodiversity within a relatively short time. Each event allowed species previously living in small ecological niches to become the most dominant species. For example, as a result of the Triassic-Jurassic event 200 million years ago that was probably caused by massive volcanic eruptions, all *archosaurs* but crocodiles and the bird-like *avemetatarsalia* from which stem all of the dinosaurs we know of became extinct.

Likewise, 65 million years ago, a massive asteroid impact near today's Mexico caused the Cretaceous-Tertiary event which caused the extinction of the dinosaurs and gave rise to mammals. Other possible causes of massive extinction events are climate change, gamma radiation bursts of nearby supernovae, or life forms like the bacteria who first produced then-toxic oxygen, or terraforming humans today.

While we see massive changes of the ecology on Earth during these events, our previous discussion still holds. The changes are *shifts* not *mutations* within the populations. Suddenly, a single rare gene might have been the crucial factor that determined survival, so whatever other genes that individual organism had would spread as well.

It is not so much that evolution made a "jump." It is simply that the environment and the factors influencing the evolutionary selection suddenly changed. This led to species on the fringes of the ecology suddenly becoming the most adapted species within their environment. These massive extinction events underscore that even under such circumstances, evolution still moves in relatively small steps.

4.2 Chemical Evolution

It is mere rubbish, thinking at present of the origin of life; one might as well think of the origin of matter.

—Charles Darwin, *Letter To J. D. Hooker*

Question

What are some specific examples in nature of complex processes or entities assembling themselves, with no other help but environmental influences?

As we have learned in the first book,[3] philosophy is hierarchical. Before we can know how we should act, we have to know what exists, how we know, and who we are. And only through understanding our ancestry, all the way back to the origin of life, can we get the full picture. In that regard, knowledge of physics, chemistry, biology, evolution, and consciousness forms the bridge between the philosophy of knowledge and the discussion of ethics in the next book.[4] We have already discussed physics, and the principles of evolution, let us now focus on the basis of life—chemistry.

---

[3]Lode, *Philosophy for Heroes: Knowledge.*
[4]Lode, *Philosophy for Heroes: Act.*

While modern life consists of cells, DNA, and complex cellular mechanisms, basic properties like replications can be easily created. But when discussing biochemistry, we have to move away from the very scientific way of looking at nature by looking at its parts. Nature is not a workshop where someone or something designs and assembles items. A better analogy for nature is: throwing all the ingredients into a pot and then waiting for them to arrange themselves. Nature does not have to deal with providing exact locations or movements for most of its operations; nature waits until the randomly moving particles are at the right place and then continues. In the same way, a filter does not individually pick and drop the small grains of sand but (assuming someone shakes the filter) simply "waits" until the larger stones and the grains are separated.

4.2.1 Cooling and Heating Cycles

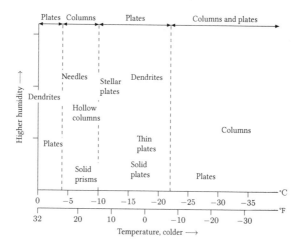

Figure 4.2: The diagram shows the range of different structures simple water can form if the environment cools down to a particular temperature.

As water cools, the water molecules arrange themselves in a low energy state: ice crystals. Depending on how the water is cooled down, the molecules have a different amount of time to reach their fixed state. The slower the cooling, the more compact the ice crystal structure becomes (see Figure 4.2). To create certain types of snowflakes, the environment needs to provide specific temperature conditions and a certain humidity in clouds, mixed with winds created by temperature variations caused by Earth's rotation and the sun (see Figure 4.3).

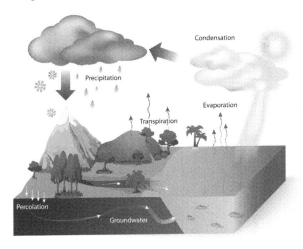

Figure 4.3: An example for a natural and non-biological cycle of water (image source: shutterstock).

The example shows what kind of complexity can arise from simple physics and chemistry without even having cells with genetic material. If an organism had to create a snowflake, it would require a complex set of genetic instructions while nature builds these structures for free by merely having the physical laws play out. The first life forms did not have to invent chemistry and physics out of the blue, they were rather an extension of them. Other structures require both heating and cooling for their assembly. The conditions for this process can be found underwater near hydrothermal vents (see Figure 4.4). Underwater volcanoes spew out hot, mineral-rich streams of smoke into the cold ocean water. Like the evaporation and precipitation of water in the air, this creates a circulation of rising hot water and falling water that was cooled down.

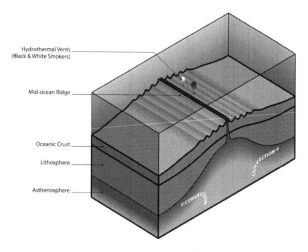

Figure 4.4: Illustration of how hydrothermal vents can form on the ocean ground (image source: shutterstock).

Figure 4.5: Illustration of a molecule forming tubes by connecting to the lower or upper side of other molecules (image source: shutterstock).

Imagine a mineral-rich stream of smoke rising upwards from an underwater volcano. Let us assume the stream contains circular molecules with the property of attaching to each other when in cool water. The hot vent ejects molecules upwards, the water slowly cools down, and molecules start to connect and form tubes. After a while, more and more molecules connect to each other into these tubes until the water with the molecules is falling down again toward the vent. This applies heat to the long tubes and they break off into smaller tubes, again attracting other freely moving molecules while rising up again with the hot stream. This process continues and more and more tubes are created (see Figure 4.5).

Based solely on the circulation of the water driven by volcanic activity, as well as molecules with a specific property of connecting in cool water and breaking off in hot water, this model shows how simple self-replication could work without having complex machinery. Obviously, when talking about the origins of life, we are thinking about cells. Also, we have not described the properties of these tube-building molecules. So, let us now discuss the actual chemical properties of certain structures.

4.2.2 Soap Bubbles

The thing most closely resembling what we know as "cells" are bubbles we see in water. Especially when mixing water with soap, bubbles form naturally. Soap bubbles are basically water bubbles with soap molecules on their inside and on their outside with the water trapped in between. These bubbles have interesting properties: they are stable for a while, they can connect with other bubbles, and you can split the bubbles into two—given enough skill. While this already sounds like a "cell," nobody claims that soap bubbles would be a good basis for life. Still, let us look closer at their properties.

$$CH_3 \underset{H_2}{\overset{H_2}{C}} \underset{H_2}{\overset{H_2}{C}} \underset{H_2}{\overset{H_2}{C}} \underset{H_2}{\overset{H_2}{C}} \underset{H_2}{\overset{H_2}{C}} \underset{H_2}{\overset{H_2}{C}} \underset{H_2}{\overset{H_2}{C}} \underset{H_2}{\overset{H_2}{C}} \overset{O}{\underset{}{C}} O^- Na^+$$

Figure 4.6: Sodium stearate, a common part of soap, with a water-attracting head (right) and a fat-attracting tail (left).

Soap consists of long molecules with two ends (see Figure 4.6), the "head" that is chemically attracted to water ("hydrophilic"), and the "tail" that repels water ("hydrophobic"). When you drop soap into water (and apply energy, meaning you stir the water), the natural reaction of the soap molecules will be to have their "head" turned toward water molecules while their tail turns away from water molecules (for example, toward bubbles of air, caused by the stirring). Like arranging triangles in a circle (or pieces of a round cake), geometrically, this leads to round objects (micelles). In moving water, the soap molecules will all be arranged to "enclose" non-water particles, for example oil droplets—you can notice this effect simply by washing your hands with soap: they get clean (see Figure 4.7).

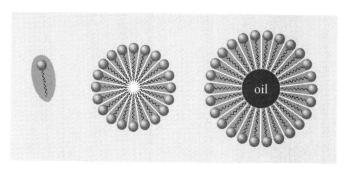

Figure 4.7: Illustration how a soap molecule encloses an oil droplet, making the oil water soluble and easy to wash away (image source: shutterstock).

4.2.3 Fatty Acids

While soap bubbles look interesting, for proper cells, we need a protective "neutral" barrier of any size—a cell membrane that divides the interior from the exterior world. The challenge is to find chemicals with the right properties. Meet fatty acids like phospholipids. As opposed to soap molecules, they have two tails, making their lower part as wide as their head (see Figure 4.8). Instead of round structures, they create flat surfaces. As opposed to micelles, they can create much larger structures that consist of two layers of phospholipids. The downside of phospholipids is that they do not naturally create bubbles like soap does, they need help from their environment in the form of pores. When you push a surface of phospholipids through a pore, spherical *liposomes* form. While modern cell membranes have a lot more structures arranged on the membrane, such as immune system markers, their integral parts are nothing but two layers of phospholipids (see Figure 4.9).

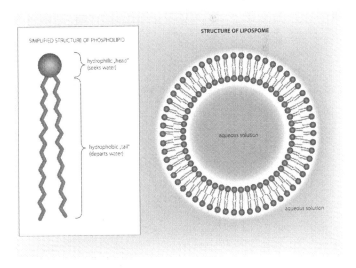

Figure 4.8: Structure of phospholipid creating a liposome (image source: shutterstock).

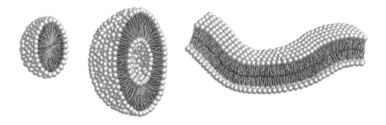

Figure 4.9: Flat surfaces, micelles, and liposomes created by fatty acids like phospholipids (image source: shutterstock).

Figure 4.10: A simple example of how bubbles are created from a flat surface are soap bubbles: blow onto a flat film of soap and you create bubbles—no special machinery except for a single "pore" needed (image source: shutterstock).

Just like phospholipid membranes can grow, these liposome bubbles can grow by integrating more fatty acids into their structures. And while the exact conditions still have to be figured out, you can imagine that very large liposomes can tear and split into two due to external forces like hitting other particles or simply heat—or by passing through the pores by which they were originally created. You can imagine cooled down minerals near the hydrothermal vent that has such pores and through which the liposomes pass again and again. Like a chemical catalyst, this mechanical catalyst helps dividing the "cell" into two. This process keeps repeating given that enough fatty acids are available in the environment and is in principle very similar to the growing tubes we have mentioned earlier.

Idea

Examples of self-assembly in nature include molecules that form longer and longer tubes until they break into two, only to grow again, as well as bubbles of fatty acids or soap that grow by incorporating more molecules into the bubble, and then divide, only to grow again.

Question

Why does the process of repeated chemical self-assembly not count as evolution?

Even with these examples in mind, the jump between lifeless matter and self-replicating life seems too great. Dividing strings of molecules or even structures like liposomes that resemble modern cells fulfills the first condition of evolution—creating copies of itself—but there is no variation. The underlying molecules—fatty acids—are provided by the environment and do not change, nor has the liposome structure any influence on the molecules. As such, liposomes still look like random, non-living products of their environment. Even if heat or radiation changes some of the fatty acids, these properties are not copied. Even if the liposome gathers some proteins along the way that help with cell division, these proteins are not copied. The property missing from these liposomes is the ability to learn.

> **Idea**
>
> While self-replicating chemical systems can be found in nature, they show no variation. They fully depend on the available material in their environment but do not drive evolution themselves.

So, when talking about variation and copying this variation to future generations of life forms, we are talking about *learned information*. Hence, we will examine the origins of life from a purely information-based point of view in the following section.

4.3 Artificial Life

> It is raining DNA outside. On the bank of the Oxford canal at the bottom of my garden is a large willow tree, and it is pumping downy seeds into the air. [...] [spreading] DNA whose coded characters spell out specific instructions for building willow trees that will shed a new generation of downy seeds. [...] It is raining instructions out there; it's raining programs; it's raining tree-growing, fluff-spreading, algorithms. That is not a metaphor, it is the plain truth. It couldn't be any plainer if it were raining floppy discs.

—Richard Dawkins, *The Blind Watchmaker*

In the previous section, we have discussed how entities could *assemble* themselves with the help of their environment, with dividing bubbles of oil being a primary candidate as a precursor for life. While the discussed liposome structures were able to self-assemble and—with the help of the environment—even replicate themselves, they were strictly dependent on their environment and had no way of actually evolving. Now, let us look at the opposite side—pure information—and combine both the chemistry and the information—the phenotype and the genotype—in the next section on the origin of life.

Example

Imagine you have a population of artificial organisms whose only task is to provide a number. To solve this problem, we will be using an evolutionary algorithm which works by applying mutation, procreation, and testing of the organisms against the environment. The environment is the fitness function which selects those organisms with the best guesses. They procreate, their offspring undergoes some mutation, and then the offspring guesses a number.

Let us say 7 is the number to guess and in the starting population, all formulas result in "1." They have all "guessed" similarly well, so they all procreate equally. In the new generation, there are now some that guess 1, some that guess 0, and some that guess 2. Those that guessed 2 procreate the most because 2 is closer to 7 and soon they overtake the whole population. This repeats in subsequent generations, with those guessing closer to 7 procreating the most. After a few generations, you will end up with a population which mostly guesses 7, with a few mutants left and right guessing more or less than 7.

4.3.1 Tierra

Of course, it is important to note that just simulating an evolutionary algorithm on the computer is not "life." One step closer to nature is the project *Tierra* by the biologist Tom Ray. In *Tierra*, the world is one-dimensional. All the organisms reside in the memory of a computer. Initially, a single organism is placed in the world. This initial organism consists of a series of instructions to copy itself. After a copy is finished, that copy is activated and starts copying itself, too. In addition, there are mutations that might improve or destroy the copying process: this fulfills all three conditions of evolution (copying, mutation, operating for its own advantage).

Figures 4.11, 4.12, 4.13, and 4.14 show an evolutionary race between hosts and parasites in a soup of the Tierra Synthetic Life program developed by Tom Ray. Each individual creature is represented by a colored bar, colors correspond to genome size.[5]

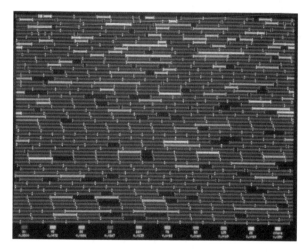

Figure 4.11: Hosts, red, are very common. Parasites, yellow, have appeared but are still rare. *Photo credit: Marc Cygnus*

---

[5]Ray, *Tierra Photoessay.*

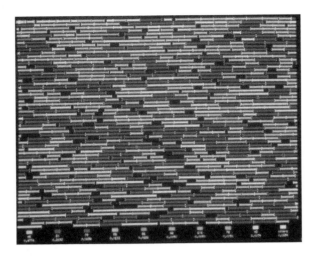

Figure 4.12: Hosts are now rare because parasites have become very common. Immune hosts, blue, have appeared but are rare. *Photo credit: Marc Cygnus*

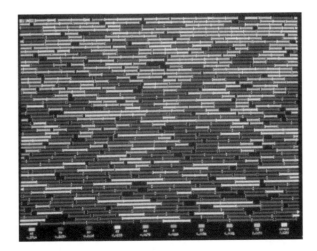

Figure 4.13: Immune hosts are increasing in frequency, separating the parasites into the top of memory. *Photo credit: Marc Cygnus*

Figure 4.14: Immune hosts now dominate memory, while parasites and susceptible hosts decline in frequency. The parasites will soon be driven to extinction. *Photo credit: Marc Cygnus*

What does the initial organism that is put into the memory look like? Here is a (simplified) description:

1. Beginning marker

2. Set first counter to address of beginning marker

3. Set second counter to address of end marker

4. Find free space in memory, save in third counter

5. Start new child organism at address of third counter

6. Loop marker

7. Copy one instruction from address of first counter to address of third counter

8. Increase first counter

9. Increase third counter

10. If first counter is smaller than the second counter, jump to Loop marker

11. Activate child

12. Jump to Beginning marker

13. End marker

The actual program is longer (80 instructions) as it uses less powerful instructions and markers that need several memory slots. But this basic setup was enough first to fill the memory with copies of the program, and then to have the programs optimize themselves: shorter programs get copied faster and eventually outrun the older, longer instruction sets.

Eventually, a mutation caused one program to change the beginning marker of a neighboring program: the first virus program emerged. The virus co-opted the copy process of other programs in the memory to have them build copies of the virus instead. Instead of going through all the 13 steps themselves, the virus looked for a host program (or found them by luck) and manipulated the second and third command to point to the beginning and end of the virus instead of the host program. This way, the host program made copies of the virus instead of itself, while the virus was able to focus all its processing time on finding hosts and manipulating the beginning and end counter commands.

This then led to a back-and-forth with programs trying to protect themselves against those parasites by hiding instructions for the beginning and end markers better, or even running a form of self-diagnosis. Later, programs developed that worked together with other programs to defend themselves and protect their code.

The setup of *Tierra* is one step closer to the origin of life as it removes one layer of abstraction: There is no separation between the phenotype (the body) and genotype (the genetic information building the body). Instead, the genetic information is directly exposed to the environment. But where *Tierra* differs from the real world is that these instructions are read and executed by *Tierra* itself. Like a computer that reads a program from the hard disk or a DVD, the "reader" and "executor" are not part of the program itself. But in real life, the genetic code is not magically turned into new cells the way it happens in *Tierra*. In nature, the genetic code is read by specialized cell machinery (which had yet to evolve on the ancient Earth) that uses it as a blueprint to build proteins.

4.3.2 Cellular Automata

So, what are other examples of self-replicating machines?

If we want to find an underlying rule—as we did with the folding or magnetic pendulum—we need to examine how these cells work. The challenge here is that the whole pattern cannot fit into the genetic code. In addition, there is no central organ but the cells themselves to coordinate the creation of the pattern. Hence, we are looking for rules that work on an individual cell. What we are looking for are cellular rules—or cellular automata as they are called in computer science.

Cellular automata are comparable to biological cells. A single cell alone does not know about the body as a whole, it can only communicate with its neighbors using chemical signals. Despite these limited abilities, if you have a lot of cells, they can work like a computer (although a lot more slowly) and produce seemingly complex results.

Example

Rule 30 is a one-dimensional binary cellular automaton rule. In this rule set, each cell can only be either in state 1 or state 0. The 30-rule says that a cell changes its state to 1 when exactly one of these conditions are met: only the cell on its left is in state 1, only the cell on its right is in state 1, only the cell itself is in state 1, or only the cell itself and the right neighbor are in state 1. In all other cases, it changes to (or remains at) state 0. When starting with a single cell in state 1, and adding a new line for each new application of the rule, the following pattern emerges (see Figure 4.15).

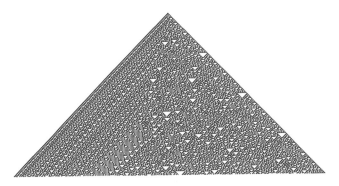

Figure 4.15: The first 256 lines of the output of the "rule 30" cellular automata.

As we can see, this very simple rule can produce a complex looking pattern when applied multiple times. Interestingly, this pattern can be found on the shell of the sea mollusc (see Figure 4.16). Similarly, cellular automata can be found for zebra stripes (see Figure 4.17). In that regard, we can determine that the pigmentation is neither complex, nor random: we can find rules that—when applied repeatedly—produce these patterns. Let us now use these cellular automata to create rules and patterns with self-replicating properties.

Figure 4.16: The shell of a mollusc showing a pattern reminiscent to the output of a cellular automata. Copyright 2005 Richard Ling.

Figure 4.17: The pigmentation pattern on a zebra (image source: shutter-stock).

4.3.3 Game of Life

Historically, the question about machinery that builds itself was first posed by John von Neumann (who was also involved in creating computers) in the 1940s. In his book *Theory of Self-reproducing Automata*, he discussed the design of machinery—a "universal constructor"—that can build copies of itself. It is a software program that runs on a computer-simulated infinite two-dimensional plane that is divided into individual, square cells, each with one of 32 states. It consists of a data tape (the genotype) and a constructor that uses the data to construct a new constructor and copy the tape.

While it showed self-replication, the inner workings were too complex to serve as a good example. A more simple and more famous example of artificial life is John Conway's *Game of Life*. It is also a *cellular automata* and loosely represents basic ideas from nature. Here, each cell can only be "alive" (1) or "dead" (0). If a dead cell has three live neighbors, it becomes alive (it is "born"). If an alive cell has two to three live neighbors, it survives. If there are more or less neighbors than that, it dies. In *Game of Life*, there are many special configurations with various outcomes, including static patterns or oscillating patterns with various intervals (see Figure 4.18).

These simple rules make the world of *Game of Life* computationally complete. That means that it is possible to run any sort of algorithm within such a system, given the right start configuration. More complex examples include digital clocks,[6] a programmable computer simulation connecting to a printer,[7] or even a simulation of *Game of Life within Game of Life itself*.[8] The idea of self-replication was later moved to the standard *Game of Life* rule set. *Gemini* is the first self-replicating structure created within *Game of Life*. It consists of a

---

[6]Stackexchange, *Build a digital clock in Conway's Game of Life*.

[7]Loizeau, *Game of life: Programmable Computer*.

[8]Bradbury, *Life in Life*.

long instruction tape that coordinates the replication, and one "arm" on either end. It takes around 33 million steps to construct a copy of itself.[9]

Still lifes	
Block	
Beehive	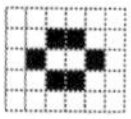
Loaf	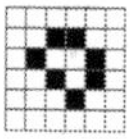
Boat	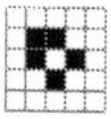
Oscillators	
Blinker (period 2)	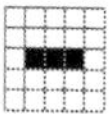
Toad (period 2)	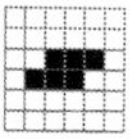
Beacon (period 2)	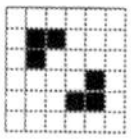
Pulsar (period 3)	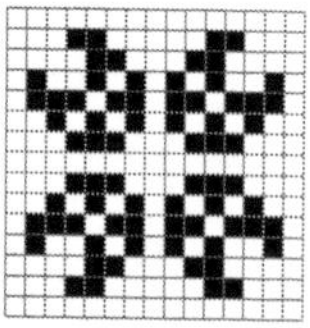
Spaceships	
Glider	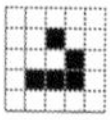
Lightwight spaceship (LWSS)	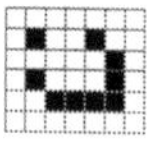

Figure 4.18: A selection of special configurations in *Game of Life*.

---

[9]Scientist, *First replicating creature spawned in life simulator.*

Given their complexity, it is hard to see how either *Gemini* or the *Universal Constructor* could evolve out of a non-living environment. The only way imaginable that they could have evolved out of a random configuration of their environment is pure chance. Given their size, the probability is basically zero. But despite both of these artificial machines looking contrived, they show how it is, in principle, possible to build a self-replicating machine with very basic rules and without centralized control. This is the situation we face when looking at chemistry: there is no "central processor" like in *Tierra*; all rules of the system work at all places, dependent only on the configuration of molecules that float around.

Despite the basic similarities, there are many differences between a *Game of Life* organism (and the *Universal Constructor* of John von Neumann) and real life forms. *Game of Life* certainly has a number of structures that look like they were "building blocks," but they are never used to build anything; they are byproducts or temporary placeholders. Strictly speaking, the "building blocks" in *Game of Life* are but "1" (black cells) and "0" (white cells). The building blocks "come alive" only by the computer interpreting them in a specific way. The individual bits that represent an organism in those worlds have no inherent properties that could affect the computer's memory.

While each example in this section can serve as a good analogy to nature, and how self-replication and even evolution works—a computer simulation allows us to watch evolution in real time without having to wait for thousands of years—none of the examples help us to explain how life *originally* developed. With this in mind, let us move on, combining the ideas of copying information with the ideas of self-assembling structures in chemistry in order to explain the origin of life.

4.4 The Origin of Life

How would you create a life-like model of your face? A simple approach is to cover your face with papier-mâché, let it dry, detach it from your face, then fill the inside of the mask with plaster, let that dry as well, and detach the papier-mâché from the plaster. The result of this process is a plaster cast of your face—a representation, not an actual *copy* of your face.

Now, this only works because you have copied but one simple layer, namely the surface. Real three-dimensional copying (including all your bones and flesh) is not possible this way. But modern life forms, from simple bacteria to multicellular organisms, can achieve that.

Question

How do modern multi-cellular life forms solve the problem of creating three-dimensional copies of themselves—their offspring?

4.4.1 The Pen That Draws Itself

When we draw, we have an idea in our head and a tool in our hand. This is similar to the prerequisites of *Tierra, Universal Constructor*, and *Gemini*, as they all have their computer code (their "idea") and their "pen" (the computer world that executes their computer code) separated. At first glance, this gives us the classic chicken and egg problem: which was first? Without the computer code, there is nothing to copy, and without a mechanism that creates copies, there is no replication.

This problem bears resemblance to what we have discussed in the first book, *Philosophy for Heroes: Knowledge*,[10] when talking about ontology and epistemology:

> [Ontology] and epistemology are simultaneous—what exists and how we know it are the foundation that starts together. And that's why the very first axiom is "Existence exists, and the act of grasping this implies there is something, and we have the faculty for being aware of it." And thereafter we shift back and forth, "We have consciousness," "*A* is *A*," "Existence is independent of consciousness," "We acquire knowledge by reason," and so on. [Ontology and epistemology] are completely intertwined.

> —Leonard Peikoff, *Understanding Objectivism*

Whatever the first life on Earth looked like, somehow, it had to be both the pen and the idea—the genetic code—at the same time. What we are looking for are molecules that contain genetic information which itself also had chemical properties that could produce building blocks for replication of this genetic information.

While the previously discussed Gemini in *Game of Life* comes closest to what we know as life—no central processor, just relying on the laws of nature or chemistry of the environment, plus self-replication—it still does not explain the origin of life. The Gemini program shows no sign of how it could develop on its own. As such, it remains a theoretical explanation with no real-life relevance. We are looking for simple structures and rules that lead to complex functionality like self-replication, not the other way around.

---

[10]Lode, *Philosophy for Heroes: Knowledge*.

Following the evolutionary idea of developments in small steps, the first life on Earth could not have had both of those two mechanisms in place simply due to chance. In *Game of Life*, a programmer manually designed the *Gemini* program, it did not develop out of itself. For the origin of life, we either also have to rely on such a designer, or we need to find a much simpler candidate for the first life form.

4.4.2 Building Blocks

To approach the problem, we can try to look at the "pen"—nature —more closely. We can examine how modern cells procreate and hope to find clues about how the first form of life did it. In modern organisms, procreation is split into two parts, namely the phenotype (the three-dimensional body) that is not cloned, and the genotype (the one-dimensional blueprint of the body) that is cloned and that builds the phenotype from the ground up.

This type of encoding three-dimensional structures into one-dimensional data was the biggest milestone in evolutionary history. It developed probably around two billion years after life first appeared on Earth. This encoding was done with a special molecule, the deoxyribonucleic acid (DNA), similar to the genetic information in *Tierra* or *Gemini* we have discussed.

DNA has two functions: creating blueprints for cell parts, and self-replication. For the genotype, the two strings unwind and chemicals dock on individual places on the string, forming the blueprint. For replication, the two strings also unwind and two new DNA strands are created in place (see Figure 4.19). DNA consists of four different building blocks—nucleobases (cytosine, guanine, adenine, and thymine)—that encode the genetic information and are fixed on two sugar-phosphate backbones. As such, DNA is a long molecule consisting of two intertwined strings.

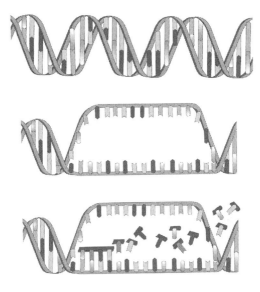

Figure 4.19: Illustration of DNA, consisting of a sugar-phosphate backbone and nucleobases. For creating partial or full copies, the DNA unwinds, allowing nucleobases to dock and to create a negative copy (image source: shutterstock).

The nucleobases are formed from amino acids, the basic building blocks of life (see Figure 4.20). Amino acids are different variations of combinations of carbon, hydrogen, oxygen, and nitrogen atoms. Geochemical research has shown that these amino acids were most likely readily available on the ancient Earth, so they are a candidate for the origin of life. The *Miller-Urey experiment* created a simple model of the ancient Earth with a mixture of water, methane, ammonia, and hydrogen as the ocean and firing electrical sparks into it (resembling lightning); 11 of the amino acids appeared in the solution.

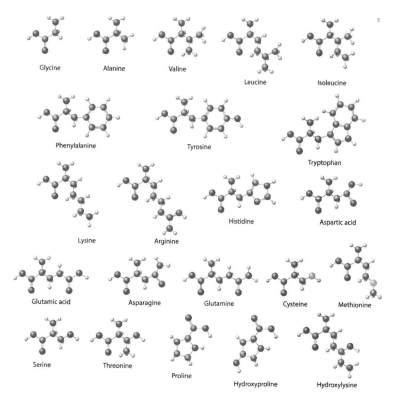

Figure 4.20: The amino acids occurring in most of modern life (image source: shutterstock).

Idea

When creating a copy of itself, with DNA, modern multi-cellular life uses the same principle as we do when making plaster casts of, for example, our hand. The difference is that, in the case of DNA, only the genetic information—not the three dimensional structure—is copied, which is only then subsequently used to build a whole copy of the original organism.

But in regard to understanding the origin of life, knowledge of highly efficient modern cells actually stands in the way. We are too inclined to approach this question by dividing it into genotype and phenotype—into design plan and organism. Where could early life have had the required machinery that allows unwinding and copying the DNA? How could it facilitate the multi-stage creation of proteins through negative copies of DNA in order to build nucleobases?

Results from DNA analysis help us to piece together the puzzle of how life evolved on Earth. We have evidence that some of the contents of our DNA go back at least 1 billion to 2 billion years, meaning that some of the same instructions that built life back then are also active now. But the challenge scientists are facing when trying to answer these questions is that there are neither survivors of any earlier life forms, nor are there any definitive traces in living organisms of the mechanisms that came before DNA. In that regard, discovering the origin of life is a detective story.

So, what we are still looking for is a "pen that draws itself"—a genotype with the ability to interact with its environment in the same way a phenotype can. But without the machinery, DNA is but information, it just rests there, without anything that would constitute a "pen": in its compact, double-stranded form, DNA has no chemical properties to speak of. On the other hand, the *blueprints* it produces for the production of proteins—the Ribonucleic Acid (RNA)—do, and it is the pen we are looking for. That is why the focus of scientists moved from the DNA world hypothesis—the idea that the first life form already had DNA—to the *RNA world hypothesis*.

4.4.3 The Conditions for Evolution

We have talked about evolution, exploring how a functioning, pro-creating organism can adapt to a changing environment over several generations. One significant point we have discovered is that it is the environment that forms the organism, not the organism pursuing its own long-term evolutionary "goals."

This idea works well until going back to the first life form who had no parents: at some point in Earth's history, there had to be a situation where at one point there was no life, and later there was. At first glance, this is a contradiction: This also meant that at one point there was not a self-replicating organism, while there was one at the next. It means that the first self-replicating life form was formed by its environment.

But even if you had a chance to take a look at the ancient Earth, you would not notice the moment when life begins—just as you will have difficulties finding the "first human," you will have difficulty finding the "first life." The first human looks very much like his or her "non-human" parents and any distinction becomes meaningless. So, the first life form was probably not revolutionary at the time. Only when looking back, could we arbitrarily set a point of life and another of "before life." Each state—even the "origin of life"—is rather something that naturally follows from the previous state.

At least, that is what scientists are trying to prove. This transforms the theory of evolution from a theory about how species evolve from one to the next, to a more general theory about life. This theory of life needs to show that starting from non-living matter, life can develop and that each step along the way can be shown to logically follow the previous one. And this is how it should be, evolution never works in jumps or miracles.

This gives us a clear goal for what we need to look for: instead of trying to come up with a scenario where all the parts magically came together, we are looking for a logical chain of small steps adding up without the need for chance. This narrows down the possibilities, making the search for the origin of life easier.

From our previous introduction into evolution, we know that at least three conditions need to be met:

1. The life form has to be able to increase the number of copies of itself in its environment.

2. A population of self-replicating life forms needs to have variations, for example caused by mutations.

3. The life form has to be able to operate for its own advantage.

4.4.4 Nucleotides of the RNA

Before discussing the RNA world hypothesis, let us first look at RNA itself. Like DNA, RNA consists of nucleobases that are attached to a strand of a sugar-phosphate backbone (forming nucleotides). When comparing DNA and RNA molecules, you can notice that RNA consists of only one strand, DNA of two. Both strands consist of a sugar-phosphate backbone with nucleobases (adenine, thymine, guanine, and cytosine for the DNA, and adenine, uracil, guanine, and cytosine for the RNA) representing the genetic information (see Figure 4.21).

DNA Structure vs RNA Structure

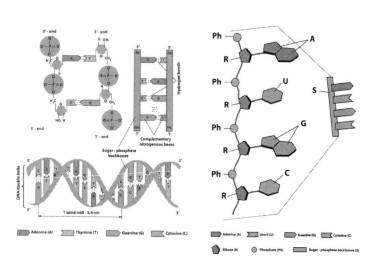

Figure 4.21: Comparison of DNA and RNA (image source: shutterstock).

But where did the nucleotides come from? In modern cells, nucleotides are produced in a complex chain of chemical reactions. But for the first life on Earth, this cellular machinery was unavailable. For the first life forms, the only alternative was that the nucleotides were already available in the water.

The surface of the ancient Earth was unlike what we have today. How Earth got all its water is still an open discussion. One theory is that ice comets hit the ancient Earth, melted, and thus distributed water on its surface. While this sounds far fetched, comets were much more common in the early solar system than today and ocean water is similar to water found in asteroids: this was possible to check because water has a unique signature, its ratio between normal water (H2O) and heavy water (H2O2).

This insight led NASA to look more closely at chemical reactions in space. In space, the temperatures are near absolute zero, while unfiltered ultraviolet radiation from the sun hits the comet. Recreating these conditions in the laboratory and using chemical compounds similar to those found in comets, scientists were able to recreate the very same nucleotides that the RNA and DNA are using.

4.4.5 The RNA World Hypothesis

Question

In order for DNA to translate its code into proteins, it needs a working cell machinery which was unavailable for the first life. At the same time, those proteins are needed in order to build copies of DNA. How does the RNA world hypothesis solve this chicken-or-egg causality dilemma?

What if life on the ancient Earth was not a piece of DNA but a single-strand piece of RNA? The core of the RNA world hypothesis states that a single-strand RNA piece floated around in a soup of nucleotides which, over time, docked with the RNA and created a negative copy of the original RNA. After the negative had been created, energy was applied to cut off the negative copy[11] which proceeded to do the same, creating a negative of itself as well—a (positive) copy of the original RNA. In that regard, the aforementioned "pen" would be the environment, providing the building blocks, and the RNA string simply waits until the environment is done with drawing, i.e., adding the molecules for a complete negative copy. This approach would fulfill the first condition of increasing copies of themselves in the environment.

---

[11]In the DNA world, special enzymes wind and unwind the DNA strands—enzymes unavailable for the first living cells.

But the second condition of evolution is that there needs to be variations. Sure, there might be different copies of RNA floating around, but they are only genotypically different. If the only difference between two organisms is that they have a different genetic code but if they do not differ in their other chemical properties, there is no evolution. Imagine a tiny RNA strand of just two bases (nucleotides), adenine (A) and guanine (G). It floats around, docks with uracil (U) and cytosine (C), creating a new RNA strand UC, which in turn docks with adenine (A) and guanine (G), creating a copy of the original RNA strand. Which of those strands is now more adapted to the environment? How do they differ?

What is missing here is the step from the genotype to the phenotype, meaning interaction between the genetic information and the environment beyond mere self-replication. In the above example, assuming that there are enough nucleotides available, an adenine-uracil RNA strand would be just as adapted to the environment as a adenine-guanine RNA strand. So, while we have self-replication of information, no evolution can take place.

In modern cells, the genotype (the DNA) produces RNA which is then used to produce proteins which are used to, for example, build nucleotides which improve the speed of replication. Instead of just waiting for molecules floating around in the environment to randomly dock with the RNA to create a copy, the cell actively takes part in creating the building blocks necessary for copying itself.

But RNA is very different than DNA. DNA always consists of two intertwined strands. Only when they unwind can they interact with the environment in a meaningful way. RNA on the other hand is a single strand—it can interact with itself. Perhaps the strongest evidence for the RNA world hypothesis is the fact that modern cells' ribosomes are folded RNA strands that also catalyze chemical reactions crucial for the cell's survival.

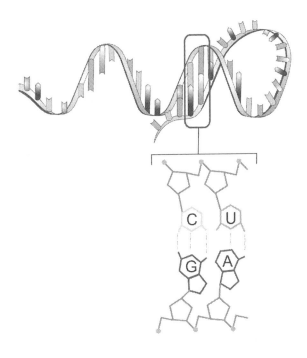

Figure 4.22: Illustration of how RNA can dock with itself, creating three-dimensional structures that can act as chemical catalysts (image source: shutterstock).

Imagine an RNA strand that has the base pairs CU (cytosine-uracil) later followed by AG (adenine-guanine). Because cytosine docks with guanine, and uracil docks with adenine, in three dimensions, it can fold into a loop (see Figure 4.22). You can imagine that with this self-interacting property, different genetic RNA codes can form any number of different structures. Each structure would have unique chemical properties, acting as catalysts for chemical reactions, and fulfilling the second condition for evolution (variety). This is something that DNA cannot do, hence there is a strong indicator for an RNA world as opposed to a DNA world.

Scientists have begun creating self-replicating RNA strands by mixing nucleotides in the laboratory. They have found that among self-replicating RNA strands, there are some that can facilitate the production of nucleotides, increasing the speed of self-replication. With this, the *RNA world hypothesis* fulfills the first and second conditions for evolution. Any change in structure that made the RNA more stable, better at procreation, and ultimately better adapted to the environment (e.g., diminishing resources) was passed on to the next generation.

While there are still some open questions, we are now very close to a full theory of the origin of life, and the ability to recreate life ourselves in the laboratory. What is now missing is the third and last condition for evolution, namely that the efforts by the RNA to, for example, catalyze the creation of nucleotides to increase self-replication, helping all RNA strands in the environment, not just the one that does all the work.

Idea

The origin of life poses the chicken-or-egg causality dilemma: which came first, the genetic code that builds the cell machinery, or the cell machinery that builds the genetic code? It is similar to the egg having to come up with a chicken that creates the egg. The RNA world hypothesis solves this dilemma by combining both parts into one RNA molecule. As opposed to DNA, RNA can interact with itself, creating three-dimensional structures that can act as chemical catalysts for building proteins. This dual nature of RNA—having both genotypical (genetic information) as well as phenotypical (catalyst for building proteins) properties—makes it an ideal candidate for the first self-replicating life form.

4.4.6 The First Cell Membrane

The third condition for evolution to take place is that the life form has to be able to operate for its own advantage. The point is that the reactions necessary to build a copy of the RNA need to happen within a protected space. Otherwise, one does the work for another organism without getting the reward. Real evolution can only happen if an organism can keep the fruits of its labor, and positive mutations are rewarded directly. From an evolutionary point of view, any chemical reactions of an organism can be hijacked by any competing organism. This is what happened to organisms in *Tierra* where viruses were able to access the interior of other organisms in the memory, stealing their energy in order to replicate themselves. It is like a group project where everyone builds parts but they all are thrown in a large bin. If someone is a lazy, he or she will just take the parts from the bin without taking part in the work.

This fundamental problem can be solved by putting the organism within a cell with a *permeable membrane.* Luckily, we have already talked about such a membrane: liposomes. All that needs to happen is that:

1. A forming liposome catches a self-replicating RNA,

2. Lets amino acids pass into the cell,

3. Has the RNA construct nucleotides from them,

4. Has the RNA replicate itself,

5. Acquires more phospholipids to grow itself, and

6. Passes through another pore to split into two cells.

Once the first self-replicating RNA strand (created from nucleotides from ice meteoroids) was captured in a liposome, the first protocell was formed. With its first procreation, evolution took its course. All evolution needed was the right mix of chemistry to provide the earliest forms of life with a whole set of tools. These tools made it possible to create structures like membranes, and even practice primitive computing: "if substance A and substance B collide, substance C is created." This allowed for the construction of chemical pathways to build more complex molecules in several steps.

Did you know?

For the scientific progress of biology, getting access to the blueprint, the genotype, allowed great advances in classification of life forms. Before this, biologists had to rely on observation and morphology. Now, they are able to calculate entire trees of relationships between all known life forms on Earth and no longer have to compare bones and take guesses. Sadly, despite humanity sharing more than 99.9% of the genetic code with each other, issues like racism persist. Science still has a long road to educate people about their true heritage, their ancestors, and how much they have deeply in common. This education—along with exposing and opposing leaders who preach antagonism, who want to divide us based on 0.1% of genetic code—is the true task of a modern hero. If we want to live in this world in peace, we need to focus on our similarities while respecting individual differences. We should be wary of the people who want to give us meaning and purpose by dividing us and creating artificial conflict. But before we can tackle such a complex subject, we will first have to find out more about our mind, and then about values and our psychology.

⟶ Read more in *Philosophy for Heroes: Act*

Did you know?

While the theory of the origin of life takes away the idea of a sudden miracle (or "chance"), it adds a deep appreciation for the universe—how it is ruled by interlocked natural laws, not by chance. Of course, *without* the knowledge of all the environmental conditions, the suggestion that life could have developed on its own indeed seems preposterous and only to be explained by something supernatural.

$\longrightarrow$ Read more in *Philosophy for Heroes: Epos*

4.5 The Red Queen Hypothesis

“Well, in our country,” said Alice, still panting a little, “you'd generally get to somewhere else—if you run very fast for a long time, as we've been doing.”

“A slow sort of country!” said the Queen. “Now, here, you see, it takes all the running you can do, to keep in the same place. If you want to get somewhere else, you must run at least twice as fast as that!”

— Through the Looking-Glass

THE RED QUEEN HYPOTHESIS · “Red Queen” is a reference the novel *Through the Looking-Glass* where the protagonist needs to run just to keep in the same place. The *Red Queen Hypothesis* refers to a stable balance between two competing species that are interlocked in their evolution and where both are evolving, yet no side gets the upper hand. They are bound to each other in a way that the failure of one species is the success of the other. A gazelle failing to outrun a cheetah ultimately helps the cheetah to procreate. Likewise, a cheetah failing to catch a gazelle helps the gazelle to procreate. Another example would be plants and caterpillars, with the plants producing more and more toxic leaves, while the caterpillars become more and more resistant to the toxin. Over time, both species become increasingly specialized in catching or evading each other.

Figure 4.23: The Red Queen advising Alice to run at least twice as fast if she wants to get somewhere (image source: shutterstock).

Genetic evolution is the tool of choice for nature to adapt to changes in the environment. For larger animals, there is an evolutionary race going on between predators and their prey: the slowest prey is killed by the fastest predators. If a mutation improves speed on either side, it puts evolutionary pressure on the other. If, for whatever reason, one side ever gained a very significant advantage, the hunt in the next generation would be so successful that only a few prey animals would be left. The predators would end up starving due to their own success and the balance would slowly return as only the very best prey survived.

For example, the speed of a gazelle cannot really compare to the speed of a cheetah, but the cheetah can hold its speed only for a very short time. So, the cheetah has to sneak close enough to the gazelle for the hunt to be successful. If the eyes of the gazelle improved, the camouflage of the cheetah would have to improve. Those gazelles that saw the cheetah approaching would survive, while those that did not would end up the victims of the cheetah. In the case of gazelles, the survivors pass on their genes to the next generation, which will be more careful, faster, or more endurant. Meanwhile, the cheetahs that are faster, better camouflaged, or more stealthy pass on their genes to the next generation. Given no significant changes in the environment, this kind of evolutionary race can go on forever. But because the lifespan of a cheetah is similar to that of a gazelle, neither side gets the upper hand; the speed of genetic changes on both sides is very similar, too.

The case of an evolutionary race is quite different when it comes to humans and the viruses that "attack" us. The generation cycle of viruses is only a few hours, compared to humans' two decades. Thus, a virus can adapt to our genetics 100,000 times faster than our genetic code can adapt to the virus. In addition, there are more than 200 different (known) viruses that can infect humans. With these odds, a more adaptive system than sexual recombination of our DNA is necessary. To understand how it is even possible that animals with such long generation cycles (including humans) could survive in the presence of viruses, we need to take a look at the basic workings of the human immune system. We will look at a simplified model of the immune system.

4.5.1 The Innate Immune System

The so-called "innate immune system" is based on our DNA and provides physical protection against pathogens. Most pathogens can never reach our organs or bloodstream because our skin and mucus (in our nose, mouth, and lungs) physically and chemically block anything from entering the body from the outside. In that regard, the human body evades an evolutionary race with the pathogen on a genetic level: it does not matter what structure the virus has, it cannot pass through barriers such as mucus. Exceptions to the rule would be sexually transmitted diseases or pathogens transmitted by insects (or generally, bites by any type of animal) that bypass mucus or skin.

The second line of defense is inflammation. If the body detects that the outer defense was breached (e.g., because of a wound), the affected area becomes inflamed. So-called "macrophages" (see Figure 4.24) call on neutrophil cells to help destroy incoming pathogens —as well as healthy cells nearby, just to be on the safe side. The neutrophil cells use a variety of strong oxidizing agents like hydrogen peroxide or hypochlorite to chemically destroy the structure of a pathogen.

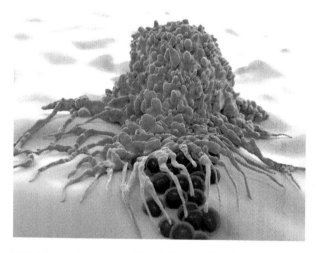

Figure 4.24: Macrophages engulf and digest cellular debris and pathogens (3D rendering, image source: shutterstock).

Beyond coordinating inflammation, macrophages have wide-ranging abilities. For our discussion, it is relevant only that they seek out pathogens (as well as dying body cells). The macrophages encounter and engulf the pathogens, break them down chemically, coordinate the neutrophil response, and present parts of the pathogen structure to the adaptive immune system.

The drawback of this system is that neutrophils come to *any* injured tissue, even internal injuries that do not have a chance of getting infected by external pathogens. This is a design flaw. The culprits are mitochondria which reside in our cells, powering chemical processes in a symbiotic relationship. There, they also hide from the immune system. But when a cell dies, the mitochondria get spilled into the bloodstream. While this sounds harmless, mitochondria being part of animal cells for billions of years, they are *still* recognized by the body as a foreign entity that has to be killed and that activates an innate immune response, including unnecessary, sometimes even chronic accompanying pain.[12] This is because mitochondria carry their own genetic information and thus do not show the same markers on their surface that would identify them as a body's own cell. Outside of a body cell, to the immune system, they look like pathogens. But despite being annoying or even torturing when it comes to chronic pain, it is a very safe approach: if something happens and cells are destroyed, from the perspective of our body it looks like something harmful must be going on, such as an injury or infection.

While it is our first and most important defense, the drawback of the innate immune system is that it is non-specific. That means that it has to somehow figure out whether something belongs in the body or not and it might also accidentally target healthy cells. And while this system is relatively fast in its initial response, it still takes some time for each pathogen.

---

[12]See Ingraham, *Why does Pain Hurt?*

4.5.2 The Adaptive Immune System

> **Question**
>
> What is a more efficient system to distinguish between friend and foe than to look at each case separately?

To answer this, let us look at the basic problem the immune system has to solve. It needs to protect us from pathogens, so the simple answer would be to attack anything that is not the "self." You could compare it with the police who look for a certain criminal. They have the fingerprints and are selective about who to take to prison and not level the whole neighborhood in the hope that doing so will also hit the criminal. Whether real or the innate, body "police" have to accurately differentiate between friend and foe. The innate immune system does not have such precision, though. While it follows a set of complex rules about how to generally differentiate between its own cells and pathogens, this process takes time—valuable time during which a pathogen could multiply. One solution to speed things up could be to give every cell the same identifier (unique for the person but identical for each cell). This would expedite detection and the immune system could simply try to destroy anything that does not carry such a marker. Using again the example of the police, it is comparable to having the exact address of a criminal versus having to control all people on the street to find the person in question.

> **Idea**
>
> Instead of spending time identifying individual pathogens, the adaptive immune system produces pathogen-specific antibodies that bind to specific pathogens.

4.5.3 A Game of Numbers

> **Question**
>
> How can complex organisms beat the evolutionary adaption speed of pathogens?

Unfortunately, the simple approach outlined above would ultimately not solve the problem. It is true that such an immune system offers an identification of "self" and would eradicate most of the pathogens invading our bodies. But a few pathogens might happen to have the same or similar markers as our cells and thus can better hide than others from our immune system. They will multiply and ultimately pose as one of our own cells after a few generations of bacterial or viral infection. And we have to remember that a pathogen evolves very rapidly: for a pathogen, procreation takes hours, while for humans, it takes 20 years from birth to maturity.

The alternative would be to give each body cell a unique identifier to make it very difficult for a virus to spread swiftly throughout your body or pose as a body cell. Unfortunately, this is impossible because it would require storing more than 10,000,000,000,000 (the estimated number of cells in our body) different identifiers in some kind of "police database" of each immune cell. Lacking such a capability, how did nature restore the balance of power in complex life forms with relatively long generation cycles?

Nature had one last trick up its sleeve. It went all in. Instead of trying to compete with pathogens by trying to identify itself, it accepts that pathogens can quickly adapt to one's cells and targets everything but the self. This also had the advantage of having a "positive" test, meaning that in the worst case a pathogen is not identified as harmful (as opposed to having body cells being destroyed because they are not recognized as innocuous). Now, that sounds like a simple thing to do, but there are around 10,000,000,000,000,000 combinations of proteins that would have to be recognized by the immune system—1,000 times more than we have cells. Likewise, the body has only around 1,000,000 proteins which could be used to recognize those 10 quadrillion proteins or protein combinations of pathogens. In an evolutionary race, we would be hopelessly lost: this number of combinations is impossible to encode into our genetic code (we have only around 3 billion base pairs, or 20,000 genes).

Instead of chaining the immune system cells to our DNA and our long generation cycles, each immune cell is different. It has a variable sequence in its DNA which results in the ability to recognize specific protein structures. Those immune cells that were successful at targeting and eliminating a pathogen were selected, and those immune cells that targeted body cells or did not target pathogens were removed from the immunological gene pool. Over time—during an infection—the immune cells become more efficient through the process of evolutionary selection. During an infection, the memory of successful immune cells is saved for possible later infections in specialized cells, providing a possibly life-long immunity.

Idea

With each immune cell being different, the immune system runs an evolutionary selection process against pathogens. This way, our immune system is able to catch up with the rate of evolutionary changes of pathogens.

This was the beginning of the evolution of a mechanism that made it possible for *evolution to happen within an organism* itself and on smaller timescales, thus outrunning the viral or bacterial evolution. What is remarkable is how a principle like evolution shows up in itself. It could be called a moment of self-reflection: it is the story of the process of evolution discovering that evolution itself is the best way to adapt to an ever-changing environment.

As with everything in nature, there was no sudden appearance of the adaptive immune system. It built upon all the mechanisms that came before. We see few remnants of the previous immune systems in our body. This is simply due to the fact that it is constantly being optimized. Every generation does it a little bit better and you never really see any jump that suddenly introduces a completely new system. The key to understanding evolution is noticing how all the intermediary steps improved the system.

HUMORAL IMMUNITY

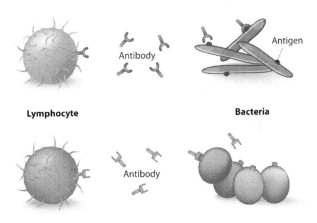

Figure 4.25: Lymphocytes build antibodies that dock to specific protein structures on a pathogen (antigen) and thus mark it for destruction for other immune cells (image source: shutterstock).

Let us now look at how, specifically, the adaptive immune system works in the body. *Please note that this is a simplified explanation. See also Figure 4.25 and 4.26 for an illustration of the process.*

1. Macrophages traverse through the body, eliminate pathogens they encounter, and present their protein structures ("antigens") on their surface (innate immune system).

2. "T-cells" with specific receptors dock with the macrophage and activate "B-cells" and present the antigen to B-cells.

3. The activated B-cells dock with both the T-cells and with the pathogen. Then, B-cells divide into memory cells as protection for later infections as well as antibody-producing plasma cells. In that regard, T-cells act as an additional layer of security to make sure that the B-cells really docked with a pathogen. Also, B-cells do not undergo deselection. Without T-cells, B-cells acting on their own would also produce antibodies against one's own body cells.

4. The plasma cells produce antibodies. Antibodies are not cells but specialized proteins that can dock with specific pathogens and can be produced in large numbers.

5. The antibodies dock with the antigens of pathogens and in this way mark the pathogens for destruction. Each antibody can identify a number of similar three-dimensional protein structures.

CELL-MEDIATED IMMUNE RESPONSE

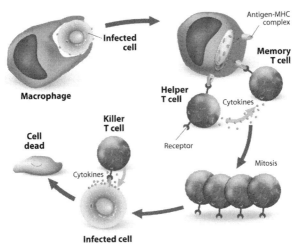

Figure 4.26: Macrophages present the antigens of pathogens to immune cells to recruit them to fight other infected cells with the same antigens (image source: shutterstock).

4.5.4 Clonal Selection

We have already noted that with the adaptive immune system, all your cells carry a unique marker. T-cells originate from stem cells in the bone marrow and are developed in the thymus to have a random configuration. That means that they can recognize different markers with a different efficiency. But before being sent into the bloodstream, and potentially mistaking your body cells as foreign pathogens, all those that actually do dock with body cells (usually around 95%) are sorted out and discarded ("deselected").

Once out in the bloodstream, a T-cell that docks successfully with a pathogen will multiply. This is the second part of the evolutionary process: antigen-specific T-cells that recognize pathogens faster will increase in number, getting faster and faster with each generation, until the adaptive immune system is so fast that it is called "immune" against a pathogen. So, we can still be infected by diseases to which we have immunity, but the adaptive immune system then works so quickly at preventing any significant spread that we do not notice.

4.5.5 Applications

Autoimmune Diseases: If something goes wrong during this dese-lection process, autoimmune diseases like diabetes, multiple sclero-sis, rheumatoid arthritis, etc. can occur. Likewise, if children do not encounter certain pathogens early in life, allergies can be the result where the immune system thinks that harmless plant or animal pro-teins are actually dangerous pathogens that deserve a full immune response. Obviously, our immune system also makes it difficult for whole foreign organs with different markers to be transplanted into the body: without additional medication to suppress the adaptive immune system, transplanted organs are rejected. Knowing how our immune system works and understanding how it originally de-veloped will help us in the future to treat autoimmune diseases.

Vaccination: If the same or a similar pathogen invades the sys-tem a second time, the body can reuse existing memory cells which can then immediately start producing antibodies specific to the pathogen. This speeds up the immune response because it saves the time of having to adapt to the specific protein structure again. This opens the door for vaccination: injecting deactivated pathogens lets the immune system create memory cells while not being in dan-ger of a pathogen actually infecting the body. As opposed to this, so-called "active immunization," alternatively, involves specific an-tibodies being injected. With "passive immunization," the body de-velops an immunity, but it is only temporary because no memory cells are created and the antibodies will sooner or later vanish.

Computer Security: For protecting computers or networks against malicious programs, a similar, albeit less complex, system is used. The system is tested at its normal "clean" state of operation and the state of the parameters are memorized by the machine. The "immune cells" of a computer security program would have to be flexible to recognize different patterns of use. For example, a website might get significantly more visitors during the day than during the night, but that should not be evaluated as a break-in attempt or "denial of service" attack. Likewise, it needs to be specific to recognize known pathways of attack, like accessing certain network addresses or trying to change sensitive parts of the operating system.

Idea

With the immune system, we have seen the power of the idea of evolution: it is so ingenious that lifeforms evolved to have within themselves mechanisms that resemble evolution. And not only can we learn more about evolution from the study of our immune system, but we can also draw analogies and parallels to security systems of any kind, be it online or in the real world. We could draw the conclusion that a system of police that takes great care of the innocent and precisely targets criminal is most effective. On the other hand, even such an advanced system as the adaptive immune system can run amok and target itself (autoimmune disease), just like an effective police force can end up being used against its own country. Our bodies have evolved to strike a balance between an innate immune system that might be too indiscriminate and an adaptive system that might turn against itself.

4.6 Evolution in the Head

> If you're good at finding the one right answer to life's mul-
> tiple-choice questions, you're *smart*. But *intelligence* is what
> you need when contemplating the leftovers in the refriger-
> ator, trying to figure out what might go with them. Or if
> trying to speak a sentence that you've never spoken before.
> As Jean Piaget used to say, intelligence is what you use when
> you don't know what to do, when all the standard answers
> are inadequate.

—William H. Calvin, *How Brains Think: Evolving Intelligence, Then
And Now*

Question

What is the main difference between the human brain and a
computer?

The main difference between a single computer and the human brain
is that you can remove a piece of a computer system and it stops
working. Just like we previously explained that adding more neurons to the brain reduces jitter, the solution in the computer sciences to improve reliability is to add redundancy by simply putting
an identical machine beside the first, to take over if the first machine
happens to malfunction. In the case of the human brain, this kind
of redundancy is found everywhere.

Example

Imagine a team of people. Each person knows his or her field very well. But when one team member gets sick, the whole project will stand still because nobody else can help out. This is the computer model: you take a piece out of the computer or a software program and it will crash or not even start. In the brain model, each team member knows a bit of everything. So even if someone is not available, the team can work together to finish a project. If you ask something, an individual might not know every detail, but together with the other team members, you can get the whole picture. In the same fashion, you can lose some brain cells and the ideas they were mapped to are still available to you, although maybe not in the same level of detail. This is the *holographic* and *redundant* model of how to manage information.

When making a decision, you can likewise imagine the workings of the brain as a committee. Individual members of this committee (neurons) could be sick at home and the rest of the committee could still arrive at a sensible decision because they share knowledge about the subject from previous meetings. That explains why even people who have lost 90% of their cortex still can make decisions like healthy people.[13]

We only "hear" the results of the committee, not the missing voices or the individual agreeing or dissenting voices. If the vote is split, introspecting to our own thoughts might show us that we are jumping back and forth to different ideas or memories all the time. How often this happens depends on our psychological makeup and the concentration of the brain transmitters. This can be compared to the "committee" changing "its mind." It constantly "votes" which thought should "pass" and become reality in the form of an action.

---

[13] cf. Calvin, *The Cerebral Code: Thinking a Thought in the Mosaics of the Mind*, p. 190.

Of course, it can take a long time for these "committees" in our mind to reach a conclusion. As long as there are several competing thought patterns, we still have to "make up our mind" and we are still indecisive about something or plan our next actions in our mind until we feel "ready." On top of deciding what to do, we also have to synchronize our thoughts.[14]

All learning involves a process of automatizing, i.e., of first acquiring knowledge by fully conscious, focused attention and observation, then of establishing mental connections which make that knowledge automatic (instantly available as a context), thus freeing man's mind to pursue further, more complex knowledge.

—Ayn Rand, *The Romantic Manifesto*

For routine tasks, we do not need this committee decision process. For example, over the years, we have become "expert walkers." Some people can easily read a book while taking a stroll. Similarly, to tie a tie or knot a hairband takes a lot of attention at the beginning. But once we have learned it, it is much easier to not try to think about it. That is also the reason why it sometimes can be difficult to explain and teach things in which we are already experts.[15]

Our "committees" only get activated when something unexpected happens, if what we perceive differs from what we expected. In that regard, the committees are strongly connected to anything related to re-examining our existing pathways of how we reach a conclusion from perception to action: learning and creativity.

---

[14] cf. Calvin, *The Cerebral Code: Thinking a Thought in the Mosaics of the Mind*, p. 169f.
[15] cf. Calvin, *The Cerebral Code: Thinking a Thought in the Mosaics of the Mind*, p. 117, p. 176.

4.6.1 Making Decisions

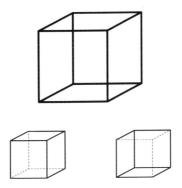

Figure 4.27: The Necker cube optical illusion showing an ambiguous line drawing. Most people see the left interpretation because people view objects more often from above, with the top side visible, than from below (image source: shutterstock).

A popular example of the workings of the committees and the "voting" is the "Necker cube" (see Figure 4.27), an optical illusion first published in 1832 by Louis Albert Necker. It is a simple wire-frame drawing of a cube that can be interpreted as either having the lower left corner or the upper right corner pointing toward the onlooker. Objectively, it is impossible to tell which way it points, simply because you see only the two-dimensional shadow of the object. But the brain is built in such a way that it always wants to "make sense" of the world. Two dimensional drawings being unknown to the brain, it assumes that we are in fact looking at a three dimensional object. Hence, your "committees" will go back and forth, with you "seeing" the Necker cube first pointing to the lower left, only to return later to point to the upper right again. An optical illusion is something where your committees think that you are looking at reality, but the image does not provide enough information to give you the full picture.

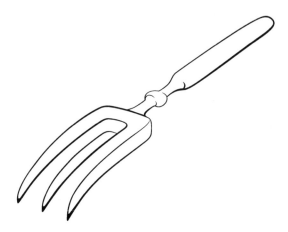

Figure 4.28: An optical illusion involving a fork with two or three tines (image source: shutterstock).

Take a close look at the fork in Figure 4.28. It has two or three tines depending on where you focus your eyes. Your mind wants to build a three-dimensional object out of the illustration but your committees are having a hard time deciding how many tines it has. Objectively, the drawing does not make sense if it is viewed as a three-dimensional object, but your subjective mind does not care. It tries to keep on "voting" which kind of three-dimensional object it is rather than consider that it might just be a drawing. We are aware of that in an abstract way, but when looking at it, we are assuming it to be three-dimensional. This kind of intuition of our mind can help us in many cases to process the world faster instead of bumping into obstacles, but we have to be aware that the model we create of the world is not necessarily how the world really is. The three-dimensional two-three-tined fork does not exist in reality, only in our mind.

Did you know?

Using optical illusions is a good way of studying the difference between one's objective perception (the ability to measure the lines on the paper to find out that it is, in fact, a two-dimensional drawing), and the subjective experience of the drawing (the creation of a three-dimensional object in our awareness). In addition, we can study how our cognitive knowledge has no influence on our subjective experience of the drawing. Sure, we know in an abstract way that it is a two-dimensional drawing, but we cannot "tell" that to our subjective experience. In *Continuum*, we are using the example only to point out the fact that given no objective way of deciding which way the cube points, our committees are going back and forth and deciding more or less randomly, or simply based on our *expectations* of how the three-dimensional object *should* look. In the next book, we will also examine other illusions, including non-optical ones where we are subjectively sure about something, but objectively the facts tell a different story.

⟶ Read more in *Philosophy for Heroes: Act*

4.6.2 The Voice of Our Cortex

While you can imagine that for a single decision, a whole number of "committees" (you!) in your brain decides which option to choose, in reality, we are often faced with an infinite number of possible decisions.

Example

While thinking about which movie to watch now, part of you also thinks about whether to grab some snacks from the kitchen, the birthday card you still need to write, the itch you become aware of only when someone else scratches his head, or the sound of dripping water from a broken faucet in the kitchen. The big question is which of those "committees" gets the say because you can do only one thing at a time.

This question becomes even more complex when we are not only faced with clear distinct options, but also when we have to sit in front of an empty sheet of paper and fill it with a painting or story without clear guidance. While it is easy for the "committees" to recognize a known pattern (like a face of a person we know), it is much harder to (re)construct that pattern or image or even invent a new one.[16]

The point of the committees is that members are in constant competition. Not only do they vote according to their own views, but they also try to convince other members of the committee and recruit them for their "cause."

Idea

The main difference between the human brain and a computer is that the brain is based on redundancy. This also affects decision making, for which there is no central processing unit like in the computer, but a series of competing decentralized "committees."

---

[16]cf. Calvin, *The Cerebral Code: Thinking a Thought in the Mosaics of the Mind*, p. 123f.

4.6.3 The Mental Landscape

Question

If every member of an idea committee votes in its own interest, all members face resistance by competition. Who determines which idea wins?

To find an answer to this question, let us take a step back. When looking at humans, we find that they show creativity and problem-solving capability unparalleled in nature. Or do they? Where is the difference between an engineer solving a physical or biochemical problem, and nature solving these problems through evolution and expressed in the DNA? How do these two seemingly totally different fields—evolution in nature and human creativity—connect?

Experiences in our life create the "landscape" of our mind. In this landscape, thought patterns compete to gain dominance. In "niches," less "popular" thoughts wait for their time of day, defending their place from other thought patterns that did not adapt to these "experience niches." It is like a jungle, home to many different "experience organisms" that come out only when we have an experience connected with it, like a familiar smell, sound, or image.

You can imagine the different thought patterns being members of different species of animal: some are hiding underground, some try to survive by climbing trees, some are courageous enough to live in the open field, etc. Each thought pattern has to try to survive and adapt in its environment consisting of other thoughts, and memories. As the vocabulary suggests, what is going on in the cortex very much resembles what is going on in nature, just at a much faster pace.

On top of memories, the current state of your brain and current sensory input influence the competition between thought patterns, too. The amount of sleep you had, drugs in your system, brain transmitters regulating your neurons, and genetics in general might promote certain thought patterns. These parameters can partially be summarized as your "personality."

Example

If you have a thought pattern in the back of your head about today's traffic, overhearing a conversation about the traffic will strengthen that thought pattern and you will become aware of an otherwise filtered out conversation. Another input is simply the environment you are currently in or the activity you are currently pursuing. Being in the supermarket, other things will "catch your eye" than when driving your car. Even the kind of music playing in the background can have a general effect on how your mind processes thoughts.

Did you know?

Certain intuitive people like to "keep their options open," and examine a lot of different ideas before they act. Their brain chemistry would decrease the speed of thought patterns to gain dominance over the others. Likewise, certain substances (like alcohol) might reduce the "vetting process" of your thought patterns so that yet unrefined thought patterns pass the barrier and enter your awareness. Along these lines, a variety of scenarios can be imagined.

⟶ Read more in *Philosophy for Heroes: Act*

> **Idea**
>
> Experiences in our life (as well as our biology and our current situation to a certain degree) create the "landscape" of our mind. This landscape influences how the committees vote and which thought pattern becomes dominant. The process is comparable to evolution in nature with competing organisms within the same geographical area.

4.6.4 Evolutionary Thought

> **Question**
>
> How do these committees and this competition work on a neuronal level?

First, the members of the "committee," the neurons themselves, do not contain concepts as such. So, you will not find, e.g., a specific "apple neuron" that lights up when you see an apple and that somehow coordinates with other neurons (also representing other things) what to do with it.

For simplicity, we can imagine concepts being represented in the brain by a group of neurons firing in a certain way. The structure of the brain allows these groups of neurons to "copy" their "role" onto other groups of neurons, making them fire the same way. If you imagine this process going on for a while, more and more groups of neurons copy themselves onto more and more neurons until most of the brain's neurons are firing the same thought pattern. This thought pattern is what "wins" the competition by sheer numbers and thus a decision is made.

This idea of a process of competing thoughts was originally devised by the neurologist William Calvin and is still part of ongoing research. As mentioned above, the point is that neurons are not computers where each one contains a single concept. Instead, concepts are always mapped onto multiple neurons. The challenge is that one group of neurons copy themselves not simply randomly around the brain, but that they keep their relative structure. With RNA, the self-replication process was done by pairing a strand with its opposite nucleotides and thus creating a negative copy which could be used to create a positive copy. With static neurons, the situation is different. Here, we want to copy impulse sequences to another neuron. How can this be implemented?

Let us first look at the basic problem of copying one image to another location on a piece of paper. For this, we can use a tool called the pantograph (see Figure 4.29). It consists of four sticks that are pair-wise connected at their center and at their ends. You fix the pantograph at one location, lead a pointer over the image you want to copy, while a pen at the end of the pantograph draws the copy.

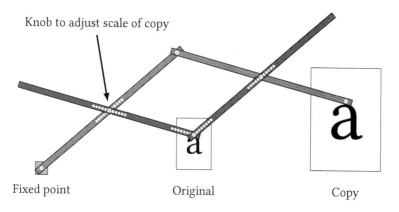

Knob to adjust scale of copy

Fixed point   Original   Copy

Figure 4.29: Copying the image of an "a" with a pantograph to another location on a piece of paper.

With neurons in the brain, you have neither a fixed point nor a pantograph. Instead, you can rely on two neurons with identical firing patterns. Their signal creates an interference pattern like we have seen in Chapter 3 with the waves. Two waves will cancel each other out or increase each other's signal strength at certain points where they meet: an interference pattern emerges. With waves, not only the location, but also the signal frequency is important. So, a group of neurons needs to send individual copies with the same group frequency. Groups of neurons in the distance can then decipher not only the signal frequency but also the spacial structure of the original group—like the spacial structure of the image of an "a" is copied by the pantograph.

Neurons that happen to be in a position where the firing patterns meet can be primed to repeat this firing pattern themselves, starting to send out the same firing pattern and subsequently recruiting other neurons as well. This way, a few neurons with an identical signal pattern can become the dominant thought pattern in a short amount of time. This constitutes the neuronal basis for natural selection within the human brain. The paper *Copying and Evolution of Neuronal Topology*[17] lays out the details of how how firing patterns can self-replicate.

Because of this copying mechanism, William Calvin calls the cortex of the brain the "Darwin machine," suggesting that an evolution of thoughts can happen.[18] And indeed it fulfills the basic ideas of evolutionary progress. Thought patterns replicate themselves, possibly with some mutations. Those thought patterns that copy themselves in the most efficient way will eventually dominate others. Likewise, thought patterns that are better suited for their "environment" (other thought patterns, memories, current sensory input, etc.) are

---

[17]Fernando, Karishma, and Szathmáry, *Copying and Evolution of Neuronal Topology.*

[18]The exact implementation pathways of the "Darwin machine" is still being examined. Current studies (Kozloski, *Closed-Loop Brain Model of Neocortical Information-Based Exchange*) show a constant loop of the signals via a number of pathways.

more "fit" to survive the mental landscape and will more easily fend off invading thought patterns. Some groups of neurons even become a very stable attractor with a lot of "guardian" attractors nearby that lead you back into the old thought patterns. To get out of these circling and repeating thoughts (neuroses), therapy is needed, for example for obsessive compulsive disorders.

The previous explanation suggests that thoughts simply are copied and then can become the dominant thought pattern. What actually happens is that, just as in nature, changes occur. Copying is not 100% exact, so new copies that might fit better (or worse) into the landscape of the brain are created. The common "I have to sleep on it" is simply an expression of letting those thought patterns run on throughout the night to make a better (or just more refined) decision. The longer we let this process run, the more refined our thoughts might become. For example, when composing music, we iterate over the process until we can say that we (all niches of our brain) are happy with the result. It "touches all the right points" and our neurons are happily firing without complaining: they do not generate new, conflicting thought patterns. The result of our thought process feels "clean."

Concerning the length of the process, this can be compared to the cooling off of materials. If we cool off a material too fast, it becomes brittle: the molecules did not have time to find their energetic minimum. If we cool it too slowly, well, it might simply take too much time. When we are tired, it becomes more difficult for neurons to keep firing in order to remain the dominant thought pattern. We are then more prone to switching between thoughts and, as a result, our concentration suffers. Likewise, if we allow a thought pattern to become dominant too quickly, the resulting ideas are also "brittle" and unrefined like metal you have thrown into water.

> ### Idea
>
> Individual neurons do not contain concepts. Instead, it is always a group of neurons that could be considered as representing a concept. The firing pattern of the group of neurons propagates through the brain in a manner similar to that of a wave. A group of neurons can recruit other neurons to repeat their fire pattern by having their waves of activation create an interference pattern in other groups of neurons.

4.6.5 Learning

> ### Question
>
> How can a neuronal network like the human brain learn?

Obviously, this copying process faces resistance as other thought patterns want to copy themselves, too. Imagine getting thousands of "cards" each day. Which ones you will react to will depend on a number of conditions, like learned behavior, other or previous signals, hormonal situation, etc. In a completely uniform brain—without memories or any learned behavior—this would lead to nothing but noise—like the noise of old analog television sets—random firing of neurons without any pattern. This is basically the experience of the world of a baby. Only over time, the neurons differ, they "learn" by getting positive feedback from the brainstem. Learning changes the brain permanently. It affects not just memories we can recall, but also how we process information. The parameters in the affected neurons are changed and they will produce different signals.

Learning for neurons is done by what is called "back propagation," which is comparable to a bucket brigade where items are transported by passing them from one (stationary) person to the next. For example, this was used to transport water before hand-pumped fire engines but it can also be seen at disaster recovery sites where machines are not available or not usable. Learning is all about reward: you want to encourage the right behavior. When fighting a fire, only the last person of the bucket brigade is actually dousing the fire. Should only that person get the reward?

In neuronal learning, indeed only the last neuron gets the reward and is encouraged for its behavior. But that neuron back propagates the reward from end to start, allowing for the whole "bucket brigade" of neurons to get rewarded and encouraged. This simple principle is also the core of artificial neuronal networks that are able to recognize and conceptualize images, or even beat human players in games such as "Go."

Example

Imagine you go to the grocery store and pick up a bar of chocolate and take a bite—your brain will be happy, back propagating the reward from the bite, to the act of unwrapping, picking up the chocolate, and walking to the store. But wait, it tastes salty! You will be very unhappy because it did not meet your expectations. Depending on the situation, this negative experience will back propagate now to a negative rating of the shop, and the brand of the chocolate. Next time visiting the shop or seeing the chocolate brand, you will have a negative feeling about it.

> **Idea**
>
> If an action led to a reward, the neurons that led to the action are "rewarded," and the probability of those neurons acting in a similar manner is increased. The neurons that led other neurons to act in this way are rewarded as well, and so on. The basic principle of neuronal learning is called "back propagation."

4.7 Outlook

In our daily lives, too often we have our creativity submerged. We can walk down the street, drive to work, bicycle, watch TV, eat, etc. without ever really using this Darwinian machine. The more accustomed we become to our environment, the more we use our automatic, non-creative mind. It also affects our perception of time: time flows more slowly for us the more we use our creative mind. To foster this creativity, it helps to walk new paths, learn new skills, move to new places, examine new ideas, and meet new people. The brain simulates what will probably happen next in order to have a decision ready. Our Darwinian machine gets activated only when our perception of reality no longer matches these predictions. Applying what we have learned about the nature of the mind will help us to improve our everyday life.

But our minds are of course more than just places for competing thoughts. We have personalities. How do they come to be? One way to approach this question is to think about how the copying process in the brain can be influenced by a variety of parameters. For example, the level of hormones and neurotransmitters in the brain can make you prefer letting this copying and evolving of thoughts go on longer than other people might, leading to more intuitive, although sometimes hesitant or even anxious behavior. Another influence could lead to more competition among thought patterns with only those best adapted to be able to copy, with the result of the person wanting to push through a particular idea as opposed to being open to a variety of viewpoints. In the next book, we will explore the connection between these basic parameters and high-level traits like our individual personalities.

With ever-increasing knowledge about our own psychology, we live in truly exciting times. But what will humanity's path look like? With the knowledge of where we came from, we can now approach the mystery of the most complex organ known to man: the human mind. Once we have solved this puzzle, we can then wonder what we should do with this information, how we can better act as leaders. Can we teach the newcomers to life on Earth—like artificial intelligences—to live sustainably and carry on the idea of life into the future?

In terms of life on Earth, *a lot* of things had to be "just right" in order for humans to evolve. We are formed by Earth's history. In that regard, we should not question evolution, but the processes (continental shifts, climate change, asteroids, the moon, ...) that led to Earth as it is today. What governs their development? How was it possible that everything came together in the right way with us as the result? Following the anthropic principle, we are able to wonder why we are here *because we are here*. If we were not here, we would not wonder. So, no matter how small the chance was for the right conditions coming together to create life and, later, to create us, it does not matter—it happened. We cannot argue that because the chance was so small, there must be another reason. The universe is so large that all kinds of situations can occur, especially if there are multiple big bangs and if the universe's age is infinite. Still, this answer *feels* unsatisfying. Beyond learning and teaching who we are, the following two books will discuss the ethical and spiritual questions of life and existence.

The Book Series
Philosophy for Heroes

 She said, "I will go no farther." "There is no choice. We can only go on." The magician said again. "We can only go on."

—Peter S. Beagle, *The Last Unicorn*

THE BOOK SERIES continues! Head over to our shop for more from this series (soon): https://www.lode.de/shop.

Part I: Knowledge. In *Philosophy for Heroes: Knowledge*, the first book in this four-book series, author Clemens Lode takes the reader on a journey, examining the foundations of knowledge. What is the basis of our understanding of the world? How does society define a "hero"? How do basic skills, such as language and mathematics, train our way of thinking and reasoning?

Part II: Continuum. Beyond the static world of the first book, *Philosophy for Heroes: Continuum* looks at gradual transitions from one condition to the next. Where do we come from? Why is there something rather than nothing? What is the source of our creativity? How can the study of natural sciences help us to understand who we are?

Part III: Act. Being a hero requires not only courage and knowledge, but also independence and consistency. *Philosophy for Heroes: Act* sets the reader's mind free from harmful manipulation by others. How can the fields of ethics and psychology help us to discover our true self, our true potential? What "masks" do people wear unknowingly? What are illusionary values? What is the meaning of life? How do we embody our values? What are the challenges we face when being independent?

Part IV: Epos. The final book in the series, *Philosophy for Heroes: Epos*, examines the influence of the most powerful tool of a leader, the *story*. Is the age-old conflict between "good" and "evil" necessary? Do heroes need "dragons"? What can we learn from the ancient stories of religion? How can we use our language for good? How can our own life become a story, an *epos*?

Recommended Reading

Beagle, Peter S. *The Last Unicorn*. Roc Trade, 1991. ISBN: 045145052-3. URL: https://amzn.to/2pYj9oW.

Bell, J.S. *Speakable and Unspeakable in Quantum Mechanics*. Cambridge University Press, 1988. ISBN: 978-05-2136-869-8. URL: https://amzn.to/2EcepkN.

Bohm, David. *Bohm—Wholeness and the Implicate Order*. Routledge Classics, 1980. ISBN: 978-04-1528-979-5. URL: https://amzn.to/2uE2k8C.

— *Quantum Implications: Essays in Honour of David Bohm*. Routledge, 1991. ISBN: 978-0415069601. URL: https://amzn.to/2ImcjB8.

Calvin, William H. *A Brief History of the Mind: From Apes to Intellect and Beyond*. Oxford University Press, 2007. ISBN: 978-01-9518-248-4. URL: https://amzn.to/2GueemQ.

— *The Cerebral Code: Thinking a Thought in the Mosaics of the Mind*. A Bradford Book, 1996. ISBN: 978-0262531542. URL: https://amzn.to/2GsrGMc.

Dawkins, Richard. *Climbing Mount Improbable*. W. W. Norton and Company, 1997. ISBN: 978-0393316827. URL: https://amzn.to/2GswpcJ.

— *The Blind Watchmaker*. W. W. Norton & Company, 2015. ISBN: 978-0393351491. URL: https://amzn.to/2H4vNe8.

— *The Selfish Gene*. Oxford University Press, 1990. ISBN: 978-01-9286-092-7. URL: https://amzn.to/2EbI0eb.

Feynman, Richard P. *The Character of Physical Law*. Modern Library, 1994. ISBN: 978-0679601272. URL: https://amzn.to/2Ilewgs.

Feynman, Richard P. and Ralph Leighton. *What Do You Care What Other People Think? Further Adventures of a Curious Character*. W W Norton, 2008. ISBN: 978-03-9332-092-3. URL: https://amzn.to/2IlxT96.

Feynman, Richard P. and Jeffrey Robbins. *The Pleasure of Finding Things Out*. Basic Books, 2005. ISBN: 978-04-6502-395-0. URL: https://amzn.to/2GqWQPQ.

Krauss, Lawrence M. *A Universe From Nothing*. Free Press, 2012. ISBN: 978-14-5162-445-8. URL: https://amzn.to/2In0GKp.

Lode, Clemens. *Philosophy for Heroes: Act*. Clemens Lode Verlag e.K., 2018. ISBN: 978-39-4558-623-5.

— *Philosophy for Heroes: Epos*. Clemens Lode Verlag e.K., 2019. ISBN: 978-39-4558-624-2.

— *Philosophy for Heroes: Knowledge*. Clemens Lode Verlag e.K., 2016. ISBN: 978-39-4558-621-1. URL: https://amzn.to/2H4wx2U.

— *Scrum Your Jira! Your Waterfall Organization Transformed Into Agile Multidisciplinary Teams*. Clemens Lode Verlag e.K., 2017. ISBN: 978-3945586655. URL: https://amzn.to/2uE3NMa.

Peikoff, Leonard. *Objectivism: The Philosophy of Ayn Rand*. Dutton, New York U.S.A., 1991. ISBN: 05-2593-380-8. URL: https://amzn.to/2q1mns6.

Rand, Ayn, Harry Binswanger, and Leonard Peikoff. *Introduction to Objectivist Epistemology*. Expanded 2nd ed. New American Library, New York, N.Y., 1990. ISBN: 04-5201-030-6. URL: https://amzn.to/2IkrMlf.

Ray, Tom. *Welcome to the Tierra home page*. [online; last accessed August 19th, 2012]. 2008. URL: http://life.ou.edu/tierra/.

Sagan, Carl. *The Demon-Haunted World: Science as a Candle in the Dark*. Ballantine Books, 1997. ISBN: 978-03-4540-946-1. URL: https://amzn.to/2IlyQOI.

The Author

" It doesn't interest me what you do for a living. I want to know what you ache for—and if you dare to dream of meeting your heart's longing. It doesn't interest me how old you are. I want to know if you will risk looking like a fool—for love— for your dreams—for the adventure of being alive.
—Oriah Mountain Dreamer, *The Invitation*

I have a passion for solving problems by applying nature-inspired algorithms. In my examination into what makes computers "smart," I found that the answer requires more than just science...My goal is to create a blueprint for heroes by providing you with a foundation in classical philosophy and modern science. My dream is to create a better world by teaching people what it means to be a hero—and how to become a leader in real life.

I studied computer science at the Karlsruhe Institute of Technology in Germany. After focusing my studies on nature-inspired optimization, and my professional career on programming, I moved on to optimizing businesses and, ultimately, helping improve individual lives.

Before creating the publishing company, I founded a company dedicated to bringing the power of nature to computers. Its most successful product was a program that used evolutionary algorithms to optimize game strategies—a task Google recently took up for its development of artificial intelligence. Right now, I am working as an Agile Coach and project manager, analyzing organization structure, team psychology, and helping with IT processes.

I would love to hear from you, just drop me a line (clemens@lode. de), or join me on Facebook (https://fb.me/ClemensLode) or Twitter (https://www.twitter.com/ClemensLode). Besides an occasional commentary on politics, and updates of my book projects, my feed is usually filled with cute animal pictures. For me, they represent innocence, opportunities, a fresh start, a positive attitude about life, and curiosity. My current interests include singing and minimalism —keeping in mind the effort and time one must invest on top of the actual price of buying a thing.

How I Used Agile to Create a Better Book

 But I should caution that if you seek to plot out all your moves before you make them—if you put your faith in slow, deliberative planning in the hopes it will spare you failure down the line—well, you're deluding yourself. For one thing, it's easier to plan derivative work—things that copy or repeat something already out there. So if your primary goal is to have a fully worked out, set-in-stone plan, you are only upping your chances of being unoriginal.

—Ed Catmull, *Creativity, Inc.: Overcoming the Unseen Forces That Stand in the Way of True Inspiration*

THE PATH OF A HERO is often a confusing one: it never leads straight to a specific goal, it is more a series of failures, a walk through a maze where sometimes it's only by taking a wrong turn that you find your way. Sometimes we have to learn the wrong thing only to understand the right thing, just as we sometimes truly value something only after it is lost.

In the process of writing a book, once the journey has started, the writer might wander off, discovering that the undiscovered path leads to places he or she did not plan to go. As such, the two possible approaches are to either change reality to fit your plan, or to change your own views. In this book series, I follow the latter idea: I let the discoveries I make along the way direct the path I will take.

In that regard, I have to admit that the *Philosophy for Heroes* series initially was planned and executed with everything but agile methodology, not even proper project management. It was an outgrowth of a hobby which turned into a book. Now that the first part is published, I have had the chance to examine the errata I received from readers and wonder, "How did that happen?!" I have shared the following curious example in a book about agile project management methodologies, *Scrum Your Jira! Your Waterfall Organization Transformed Into Agile Multidisciplinary Teams*:

Did Bilbo Sail to the West?

Was it Bilbo who sailed to the West? Reading *Philosophy for Heroes: Knowledge*, this seems to be the case. On page 5:

> **Did you know?**
>
>
>
> In the book series *Lord of the Rings* by J. R. R. Tolkien, the hero Bilbo undergoes a long journey to destroy evil once and for all. Through magical explanations, it is assumed that evil will not return once a magic ring is destroyed. In the story, after the heroic deed is done, the "resolution" for the hero is to sail away to another country and spend the rest of his life there. This idea of a final resolution of a problem is the traditional way of portraying a hero. But in reality, the real task would be only beginning: One would have to ask, how did the people turn evil in the first place? How can we educate them to prevent a similar disaster in the future?
>
>   ⟶ Read more in *Philosophy for Heroes: Epos*

Now, while the case could be made that Bilbo underwent a heroic transformation, that he fought evil, that he traveled to the West and might have used a boat at some point, the story sounds much more like Frodo's story in *Lord of the Rings*. Myself, I am well versed in fantasy literature, and the amount of media related to the subject is anything but sparse. How could such an error happen?

My usual judgment on products (in this case, a book) is that they are a mirror of the company behind them. If you have a little bit of background information, reviewing a product can be like an archaeological dig. *Philosophy for Heroes: Knowledge* is a multi-layered book. First, it is part of a series. When writing the first book, the other three books had to be kept in mind. In addition, especially being the first book, it had to stand on its own despite its dealing with the basics (philosophy and language). You cannot sell a book called *Philosophy for Heroes* and then tell the reader to wait for Part IV to finally read about what heroism means. Second, it contains a variety of components: study questions, ideas summarizing a section, biographies adding a human element to sometimes abstract explanations, and real life examples. Skimming through the book, those components seem to be "added features" that—while adding value to the book—could just as well be removed. This points to an evolution of the book. Looking back, this is actually true, it underwent a number of transformations:

1. A single, very large book

2. A five-part series

3. Then, a four-part series

4. Then, a four-part series with the first book required to stand on its own

5. Finally, a four-part series, the first book standing on its own, and additional components (study questions, ideas, biographies, examples, etc.)

As this evolution played out, the later changes underwent the least amount of review. Certain parts, that were already finished when devising the initial large book, had so many reviews, the time spent on dragging them along seemed like a waste. How does one write a book without having such a large variance of quality between its parts?

For this, we look at software development. A piece of software faces the same problem: it evolves, some parts are "fresh," others have been looked at and tested for years. The solution people came up with is called "Agile" (with one variant being "Scrum").

The best approach to writing something—anything—is to make sure that its pieces stand for themselves. The advantage of this approach is to have those pieces complete and ready, and you can publish each to get feedback and build an audience. Looking back, I should have published each section of the book separately. Sure, someone could piece all the sections together and then have a copy of the book for free. But that takes a lot of effort. Even if it is just half an hour of work, in that time one could have easily bought the book. Also, the final edit of a book surely connects the independent parts to a greater whole.

In any case, if I had followed the Agile approach, it would have been Frodo, not Bilbo, throwing the ring into the fire and traveling to the West.

Lesson learned.

Essential Steps in Implementing Agile

    The Agile movement provides alternatives to traditional project management. Agile approaches help teams respond to unpredictability with incremental, iterative work cadences and empirical feedback. Agilists propose alternatives to Waterfall, or traditional sequential development.

                                                  — *The Agile Movement* (edited)[1]

    Scrum is an Agile software development model based on multiple small teams working in an intensive and interdependent manner. The term is named for the scrum (or scrummage) formation in rugby, which is used to restart the game after an event that causes play to stop, such as an infringement.

                                                    — *What is Scrum?*[2]

[1] AgileMethodology, *The Agile Movement*.
[2] TechTarget, *What is Scrum?*

When clients ask me to help with implementation of Agile techniques by using Scrum, my first question is: "What do you mean by 'Scrum'?" Usually, I then hear that the company has its own special version of Scrum (or other Agile technique) because, according to the people with whom I'm meeting, their company is a special case.

First, yes, your company is a special case. Each company is unique and in a special market niche. Second, if you followed the evolutionary approach of improvements in small increments, you have adapted your process to the environment of the company. No two companies are alike. Hence your need for an external consultant to examine the special conditions in your company.

Introducing (and running) Scrum means that you want to change your company according to proven methods. You can't have your cake (Scrum) and eat it, too (changing Scrum to suit your company) —you can't improve your company by adapting the Scrum process to your company.

As a consequence, the reality I see all too often is that the company hires a Scrum Master who merely acts as a supporting firefighter, accompanying the former project manager (now "product owner") and running around the company putting out fires. This kind of extra resource is justified to upper management by pointing to "Agile" and its use in other companies...

> **WATERFALL** · *Waterfall* is a project management method where a product moves through a number of phases before a final version is finished for release. Compared to Agile, the problem with this method is that it requires additional communication channels between the individual phases and that the time until a team or company gets feedback from a customer is generally much longer.

SCRUM · *Scrum* is a set of management tools that focuses a project back on the team level and uncovers internal and external impediments of the production process. By reducing communication paths through small, multidisciplinary teams, as well as frequent releases to the customer for review, the probability for project success can be improved even if the scope is not clear from the start. In addition, work is divided into units of fixed lengths (sprints) which helps to plan future sprints with your team working at a sustainable speed.

SPRINT · A *sprint* is a timespan of one to four weeks within which a certain selection of stories should be finished by the team. Given the fact that the whole team spends 10 percent of the time (depending on the sprint length) planning and reviewing each sprint, the goal is to reach 100 percent completion of all stories while meeting the project's quality standards and without overtime. Like a marathon runner needs to carefully plan her energy, planning a sprint requires very good estimation skills by the teams.

SCRUM MASTER · The *Scrum Master* controls the Scrum process. Besides proactively identifying and removing impediments to the process, the Scrum Master also supports the team in meetings as a moderator and individually in personal talks. The Scrum Master also stands up against outside influence on the process, ideally by propagating the Agile idea throughout the organizations and by explaining why certain restrictions are necessary for the overall project success.

PRODUCT OWNER · The *product owner* is part of the Scrum team and represents the stakeholders. The main task is stakeholder management as well as having a deep understanding of what the project is about and being able to make decisions. A product owner fills and prioritizes the backlog, keeping the complexity estimations of the team in mind. The product owner should have full authority and the final say about the prioritization of the backlog. During the sprint, the product owner answers questions from the team about the scope of the project, as well as gives feedback about finished (but not necessarily done!) tasks, but otherwise does not interfere in how the team manages its work.

Looking at the Actual Causes of Problems

One of the techniques used in project management is to find the cause of an issue. Digging deeper, my next set of questions to the client usually focuses on the greater picture or vision of the company. Instead of telling me about their mission, they typically respond that they want to "test" Agile and then implement it in other parts of the company.

Besides noting the obvious misunderstanding of Agile as the new (local!) management technique, questions arise: What would success look like? What would failure look like? What are the concrete, measurable business objectives of the project of introducing Agile?

I am convinced that introducing Agile itself should be managed with modern project management techniques, PMBOK being my favorite. Managing Agile goes far beyond the scope of this chapter, but you certainly must have an idea about where you are going with it and what you want to achieve.

PMBOK® · *PMBOK* stands for *Project Management Body of Knowledge* and describes a generic system of workflows within a project. While it is mainly applied to Waterfall projects, many of its parts can also be used in an Agile project, like defining how the team communicates with the outside world, defining the vision and scope of the project, or defining why one would want to use Scrum at all. (PMI, *A Guide to the Project Management Body of Knowledge - Fifth Edition*)

To illustrate this further, I recommend reading Ayn Rand's introduction to philosophy that looks at the example of an astronaut stranded on a planet:

> Suppose that you are an astronaut whose spaceship loses control and crashes on an unknown planet. When you regain consciousness and find that you are not badly hurt, the first three questions on your mind would be: Where am I? How can I find out? What should I do?
>
> You see unfamiliar vegetation outside, and there is air to breathe; the sunlight seems paler than you remember it and colder. You turn to look at the sky, but stop. You are struck by a sudden feeling: if you don't look, you won't have to know that you are, perhaps, too far from Earth and no return is possible. So long as you don't know it, you are free to believe what you wish—and you experience a foggy, pleasant, but somehow guilty, kind of hope.
>
> You turn to your instruments: they may be damaged, you don't know how seriously. But you stop, struck by a sudden fear: how can you trust these instruments? How can you be sure that they won't mislead you? How can you know whether they will work in a different world? You turn away from the instruments.
>
> Now you begin to wonder why you have no desire to do anything. It seems so much safer just to wait for something to turn up somehow; it is better, you tell yourself, not to rock the spaceship. Far in the distance, you see some sort of living creatures approaching; you don't know whether they are human, but they walk on two feet. They, you decide, will tell you what to do.
>
> You are never heard from again.

This is fantasy, you say? You would not act like that and no astronaut ever would? Perhaps not. But this is the way most men live their lives, here, on Earth.

—Ayn Rand, *Address to the Graduating Class of the United States Military Academy at West Point New York* (adapted)[3]

In terms of a company, your immediate goal is of course to survive the next month. But then, you have to establish where you are on the map. You have to open your eyes, look at the sky, and check your instruments.

In terms of Agile, I recommend to my clients that they run it like a project. We know what works from hundreds of studies, and we can create a list of items that are implemented in Scrum. In that list, we simply mark the current state of the process. Often, even in non-Agile companies, some processes have already been implemented because the people who are managing projects notice which processes work. A simple approach is to check against the *Principles Behind the Agile Manifesto*[4]:

- Welcome changing requirements.
- Trust and motivate individuals on your team and other teams in the company.
- Developers and non-developers (e.g., marketers, salespeople) must work together daily.
- Face-to-face is the most efficient and effective method of getting things done.
- Progress is measured in terms of working software.

---

[3]Rand, *Address to the Graduating Class of the United States Military Academy at West Point New York*.

[4]Beck, *Principles Behind the Agile Manifesto*.

- The entire team must promote sustainable development, they should be able to maintain a constant pace indefinitely.

- The entire team must work together to continuously improve technical excellence and to enhance agility.

- Keep in mind that simplicity is valuable; simplicity is the art of maximizing the amount of work not done.

- The best solutions emerge from self-organizing teams.

- Effective teams reflect regularly on how to become more effective.

- An Agile company satisfies customers through early, frequent, and continuous delivery.

Together with the client, for each point, I detail how this is implemented in the company—this is the first step of documenting the current process. If I can't explain how it is implemented or if we find that it is not implemented, I focus on those points and try to find explanations for each: why the company is not able or not willing to fulfill this part of the Agile process. And I do not just ask, "Why?" I ask, "Why why why why why...?" until I find out the actual reasons something has not been implemented.[5]

And at this point, the real work starts: addressing those issues that hinder the Agile process on a daily basis.

Lessons Applied

With the mistakes of *Philosophy for Heroes: Knowledge* in mind, this book was created as a series of regular articles, to be released on our website. This kept each section limited in complexity and scope,

---

[5]Ohno, *Ask why five times about every matter.*

kept me and my editor focused on the task, kept any perfectionism in check, and most importantly, we got feedback. After both chapters were written and released, we combined them into a whole book, and re-edited them to fit together. I decided that it is OK to release articles to the web as early as possible. Early adopters can read them online for free, and anyone who wanted a polished version could wait for the final book. Another great advantage we found is that global design issues—like capitalization of certain words, adding a glossary, and generally streamlining the technical terms—could be tackled later. Lastly, it reduced the urgency toward the end. The chapters were already online and "working" in terms of garnering interest. People got curious about the book while we were still writing it. That, I think, is the power of Agile applied. To summarize, we incorporated:

- twice-weekly chats
- early releases for feedback
- keep perfectionism in check
- keep complexity in check
- market while still writing
- focus on global issues later

And, perhaps the most important point, enjoying a feeling of accomplishment.

Of course, this did not always work as planned, given some of the complexity of the subjects and my own tendency to write more when I really like a certain topic. But compared to the time we needed for the first book, it was a huge improvement. And if applied more consistently, the third book will be even better and be completed more quickly. Ultimately, of course, it is up to you to judge: *which of the first two books did you like more?*

Reflection

D ID YOU FIND all the *Eureka!*? Join the vibrant conversation with other readers in our online forum at: https://www.lode.de/study/pfh2

Physics

• Why should one state follow from another at all? Why is everything not "frozen?" Why do entities act according to their properties all the time? Could they not act in one year like this, and in another year like that?

• Why are ontology and epistemology simultaneous?

• What situation led to the first documented application of methodical scientific research?

• How can we determine if a medical treatment was effective, not just the body healing itself?

• What major problems stood in the way of early researchers before the scientific method?

• What was new about the scientific method?

• Given that for a phenomenon, explanations of arbitrary complexity can be made, which view of the world should we choose?

• Science has become synonymous for "truthful." This has led people to try to emulate science without following the scientific method. Some have even started calling themselves "scientists" and used their titles to give their statements more meaning. Likewise, expressions such as "pseudoscience" have become an insult. What is the role of the word "science" in daily life?

• How can we actually measure the position of a particle?

• Is light a wave or a particle?

• How can a quantum computer be significantly faster than a conventional digital computer?

• What is the ontological view of the Copenhagen interpretation?

• Why could the many worlds interpretation be called ontologically wasteful?

• What sets the hidden variables interpretation apart from the Copenhagen interpretation of quantum mechanics?

• How can structure emerge from chaos? Why isn't everything just "noise"? And how does "complexity" differ from "chaos"?

• If the "big bang" created the universe, and if nothing existed before the big bang, where did the big bang come from?

Evolution

- What are the three major properties a system needs to have in order to evolve?

- If it is all about variety, why do we still intuitively think that mutations are the driver of change and evolution?

- Are larger mutations ("macro-evolution") that suddenly give an organism a significant advantage an argument against the idea that variety is the driver of evolution?

- What are some specific examples in nature of complex processes or entities assembling themselves, with no other help but environmental influences?

- Why does the process of repeated chemical self-assembly not count as evolution?

- How do modern multi-cellular life forms solve the problem of creating three-dimensional copies of themselves—their offspring?

- In order for DNA to translate its code into proteins, it needs a working cell machinery which was unavailable for the first life. At the same time, those proteins are needed in order to build copies of DNA. How does the RNA world hypothesis solve this chicken-or-egg causality dilemma?

- What is a more efficient system to distinguish between friend and foe than to look at each case separately?

- How can complex organisms beat the evolutionary adaption speed of pathogens?

- What is the main difference between the human brain and a computer?

- If every member of an idea committee votes in its own interest, all members face resistance by competition. Who determines which idea wins?

- How do these committees and this competition work on a neuronal level?

- How can a neuronal network like the human brain learn?

Eureka!

Physics

- If everything were random, something like the human brain could not exist. Such complex structures can emerge only when there is some sort of pattern in the laws of nature. So, the question is not whether the universe is deterministic or not, but how "indeterministic" (if at all) it is. Evidence shows that the degree of "indeterminism" or jiggering in our universe is relatively small. That we exist is proof that stable systems like our bodies can form despite the jiggering at the particle level. The question remains whether we live in an indeterministic world that allows a certain level of order at the macroscopic level (biology), or if we live in a fully deterministic world.

- Ontology and epistemology are simultaneous—what exists and how we know it form a foundation of philosophy.

- Large-scale war with mass-produced weaponry led to similar injuries on many battlefields. This allowed systematic medical research with different treatments.

- A method based on trial and error is difficult to apply to future situations if there is no comprehension of what actually worked. In medicine, because of the placebo effect, even without knowing what worked, a positive belief alone can produce results. Likewise, without such a belief, even a comprehension of what worked can lead to a treatment failure.

- Before the scientific method, research was limited by bias, a reliance on supernatural explanations and premature conclusions, the absence of a scientific community, and the lack of willingness to admit failures.

- What was new about the scientific method was that this process was formalized and enhanced by peer reviews from a scientific community in order to reduce bias. Not only can your current peers test your conclusions, but also all future scientists can take your paper from the archives and retest the assumptions. Proper documentation also includes proper citations. A study whose results you have used might turn out to be erroneous. If you have properly cited that study, other scientists can try to correct your work more easily. This openness—having the courage to admit mistakes—is the real driver of scientific progress. Instead of building a knowledge hierarchy of things that are possible to prove, science constructs a knowledge hierarchy where each part openly states that it might be wrong if certain conditions are met. This way, any scientific theory can be proven false by experiment, so any theory building upon other theories always carries with it a whole tree of falsifiable experiments as a prerequisite. In that regard, the start of rigorous application of the scientific method was as significant as the invention of writing. While researchers published books before, there was no process in place to organize this knowledge or access it in a structured manner, like tracing the references back to the original study. With the scientific method, scientists were able to efficiently organize knowledge, and to *trust* and build upon the results of other scientists for the first time in history.

• For any given data, there can always be different models. Occam's Razor is useful when deciding which of the models is preferable by ranking them by their complexity.

• The troubling issue of modern science is that some scientists see themselves no longer as part of a branch of philosophy, but as philosophers themselves. As a leader, it is not enough to just remain in one field of study, trying to subjugate all other fields of study. Your task is to reach out to other fields and unite them, to go back and forth between the "what is" and "how do we know."

• Particles are in fact not tiny (classical) billiard balls. Measuring a particle's position is only possible by influencing the particle, making it impossible to know its original position.

• Light has properties of both waves and particles. It causes interference patterns like waves, but also comes in packages ("quanta"), hence the name "quantum theory" was born.

• Quantum computing is so much more powerful than conventional computing because the way the waves interfere with each other is "computed" by nature for free. Basically, quantum computers repeatedly run small physics experiments, which makes solving problems of a similar nature very efficient. The difference, compared to a conventional computer, is that the interference pattern emerges in a few steps depending on the accuracy you want. Thus, the calculations of a quantum computer are much less problem-size dependent.

• In the ontology of the Copenhagen interpretation, the world is based on pure mathematics. Physicists stay away from the question where these probabilities originate. As the Copenhagen interpretation is based on the (mathematically correct) quantum theory, its predictions are the same as any other interpretation—but it cannot *explain* anything.

• Instead of saying that the quantum world is based on probabilities, in the many worlds interpretation, it is argued that the universe is fully deterministic—insofar as it has a definite history—but that at each "moment," the universe splits into an infinite number of copies, each representing one choice among the probabilities. It would be ontologically wasteful as it requires the creation of an infinite number of universes. In addition, you would still not know *why* a certain branch was chosen.

• Scientists have found it difficult to understand the Copenhagen interpretation as it does not explain where the probabilities come from. The hidden variables interpretation replaces the probabilities with a fixed "hidden" cause, the so-called pilot wave. It might be hard or impossible to measure, but they are (ontologically) there. As such, the hidden variables interpretation is deterministic—as opposed to the Copenhagen interpretation.

• Chaos theory explains the complexity in nature by pointing out that it is the result of repeatedly applied (simple) rules or natural laws. Stable elements within a chaotic system are called attractors. They are the result of a looping repeatedly applied rule.

• What we call "nothingness" is still a region of space where the laws of physics apply. And in physics, nothingness is not stable. New particles are generated and destroyed at all times. It is conceivable that our "big bang" happened by accident some place within an infinite universe.

Evolution

• In order for evolution to occur, the life form has to be able to increase the number of copies of itself in its environment. In addition, the resulting population of life forms needs to have variation, and each life form has to be able to operate for its own advantage.

• It appears that mutations are the drivers of change and evolution, because we are looking at our evolutionary history as survivors. Only if we ignore all the attempts of nature that did not lead to a successful organism, life looks as if it is driven by nothing but a series of miracles.

• Mutations that led to larger evolutionary jumps were always the result of many smaller changes. For example, the *most recent* mutation that enabled human language was prepared for by many other genetic changes (conscious breathing, descended larynx, etc.) with their own individual advantages.

• Examples of self-assembly in nature include molecules that form longer and longer tubes until they break into two, only to grow again, as well as bubbles of fatty acids or soap that grow by incorporating more molecules into the bubble, and then divide, only to grow again.

• While self-replicating chemical systems can be found in nature, they show no variation. They fully depend on the available material in their environment but do not drive evolution themselves.

• When creating a copy of itself, with DNA, modern multi-cellular life uses the same principle as we do when making plaster casts of, for example, our hand. The difference is that, in the case of DNA, only the genetic information—not the three dimensional structure—is copied, which is only then subsequently used to build a whole copy of the original organism.

• The origin of life poses the chicken-or-egg causality dilemma: which came first, the genetic code that builds the cell machinery, or the cell machinery that builds the genetic code? It is similar to the egg having to come up with a chicken that creates the egg. The RNA world hypothesis solves this dilemma by combining both parts into one RNA molecule. As opposed to DNA, RNA can interact with itself, creating three-dimensional structures that can act as chemical catalysts for building proteins. This dual nature of RNA—having both genotypical (genetic information) as well as phenotypical (catalyst for building proteins) properties—makes it an ideal candidate for the first self-replicating life form.

• Instead of spending time identifying individual pathogens, the adaptive immune system produces pathogen-specific antibodies that bind to specific pathogens.

• With each immune cell being different, the immune system runs an evolutionary selection process against pathogens. This way, our immune system is able to catch up with the rate of evolutionary changes of pathogens.

• With the immune system, we have seen the power of the idea of evolution: it is so ingenious that lifeforms evolved to have within themselves mechanisms that resemble evolution. And not only can we learn more about evolution from the study of our immune system, but we can also draw analogies and parallels to security systems of any kind, be it online or in the real world. We could draw the conclusion that a system of police that takes great care of the innocent and precisely targets criminal is most effective. On the other hand, even such an advanced system as the adaptive immune system can run amok and target itself (autoimmune disease), just like an effective police force can end up being used against its own country. Our bodies have evolved to strike a balance between an innate immune system that might be too indiscriminate and an adaptive system that might turn against itself.

• The main difference between the human brain and a computer is that the brain is based on redundancy. This also affects decision making, for which there is no central processing unit like in the computer, but a series of competing decentralized "committees."

• Experiences in our life (as well as our biology and our current situation to a certain degree) create the "landscape" of our mind. This landscape influences how the committees vote and which thought pattern becomes dominant. The process is comparable to evolution in nature with competing organisms within the same geographical area.

• Individual neurons do not contain concepts. Instead, it is always a group of neurons that could be considered as representing a concept. The firing pattern of the group of neurons propagates through the brain in a manner similar to that of a wave. A group of neurons can recruit other neurons to repeat their fire pattern by having their waves of activation create an interference pattern in other groups of neurons.

• If an action led to a reward, the neurons that led to the action are "rewarded," and the probability of those neurons acting in a similar manner is increased. The neurons that led other neurons to act in this way are rewarded as well, and so on. The basic principle of neuronal learning is called "back propagation."

Glossary

A

Aggregate • An *aggregate* is a number of entities that have a reciprocal effect on one another, so that they can be considered collectively as their own entity (e.g., a cup full of water—all water molecules interact with each other).

Anthropic principle • The *anthropic principle* is the consideration of how the environment and natural laws just happen to support human life: only the inhabitants of those worlds that can sustain intelligent life can wonder why their own world happens to support intelligent life. If the conditions for intelligence were not met, there would be nobody wondering about it.

Argument from Authority • Just because there is an authority, be it a priest, politician, or scientist, who makes a statement, this does say anything about how truthful that statement is. While the position or title of a person certainly means that the person has been tested and has something to lose, all new statements have to be proven. Usually, a title means nothing but the proof that someone was serious about getting it, and is serious about keeping it. The quality of a title then ultimately depends not on its length, but on the community behind the title that checks its quality. Experts can lie or make honest mistakes; relying on someone's title does not protect you from that (cf. Cialdini, *Influence—The Psychology of Persuasion*, p. 208–36).

Attractor • Repeatedly applied rules or laws can eventually loop. In chaos theory, this kind of a loop is called an "*attractor.*"

C

Cargo cult • A *cargo cult* refers to the behavior where someone tries to imitate certain aspects of another (successful) person, expecting the same success. For example, celebrities are often on TV but just by managing to get yourself on TV, you will not necessarily become a celebrity.

Causality • *Causality* refers to the effect of one or several entities on another entity in a certain situation (e.g., an accident is no random occurrence, there are one or several causes which led to the accident, such as lack of sleep, a technical defect, poor visibility, etc.).

Chance • The cause of an effect when no other cause could be determined.

Chaos theory • The *chaos theory* states that small differences in initial conditions can yield widely diverging outcomes. For example, given enough repetitions, the effect of a butterfly flapping its wings on one side of the world might cause a hurricane on the other side of the world. Beyond this butterfly effect, the chaos theory also deals with patterns that emerge from an apparently chaotic system.

Complex Argument Fallacy • It is a *Complex Argument Fallacy* if you use a more complex argument (that leads to the same conclusions as the more simple one) to argue against an existing argument. The new argument requires more assumptions to be true. An extension of this argument is to simply declare a problem or situation as "complex" while dismissing simpler solutions. For example, a child could declare that her situation is "complex," complaining she is running out of pocket money, while the reasons are very clear (her expenses). This way, the chance to even discuss a subject is negated instead of addressing actual issues. This fallacy is also often used in connection with an argument from authority, implying that only "experts" are allowed to have an opinion on the subject. Of course, sometimes, there are no simple solutions and a problem is complex, requiring experts to discuss. That is why the complex argument fallacy is a reminder to reflect before either calling a situation too complex, or trying to apply basic solutions that are too simple.

Configuration of a property • The *configuration of a property* relates to the intensity of a certain property of an entity.

E

Effect • An *effect* is the change caused to the configuration of the properties of an entity (e.g., the heating of water changes its temperature).

Entity • An *entity* is a "thing" with properties (an identity). For example, a plant produces oxygen, a stone has a hard surface, etc.).

Evolution • *Evolution* is the combination of the process of selection together with a system of cloning or procreation.

Exaptation • *Exaptation* is the use of a certain trait for a problem or environment other than what it was originally "intended" for. An example would be feathers that started out as heat insulation, and only later were used to improve jumping, and finally for flight.

F

Fitness landscape • The *fitness landscape* is the sum of all environmental influences on an entity. For example, if you are sifting sand, a riddle screen lets small particles of sand fall through while larger stones are retained. In this case, the riddle screen and the shaking of the riddle screen would be the "fitness landscape." In nature, the fitness landscape would simply be the environment over time, including all other life forms, the climate, etc.

Fractal • A *fractal* is a self-similar pattern created through repeatedly applying the same rule on itself.

G

Generation • A *generation* is a set of systems during one cycle of procreation.

Genotype • The *genotype* is a system that is the blueprint for the phenotype.

I

Identity • An *identity* is the sum total of all properties of an entity (e.g., weight: 160 pounds, length: 6 feet, has a consciousness, etc.).

M

Mutation • A *mutation* is a change of the genotype of a life form. This change can, but does not necessarily, have consequences for the phenotype.

O

Occam's Razor principle • According to the *Occam's Razor principle*, among competing hypotheses, the one with the fewest assumptions should be selected. It is attributed to the English Franciscan friar, scholastic philosopher, and theologian William of Ockham (1287 - 1347).

P

Phenotype • The *phenotype* is the actual body of a life form. Changes in the phenotype generally do not have effects on the genotype. Generally only the phenotype interacts directly with the environment.

Process • A *process* describes the mechanism of a cause working to an effect (e.g., if you put an ice cube into a glass of water, the cooling of the water is the process).

Procreation • *Procreation* is a process by which a system creates (on its own or with the help of the environment) a new entity with a similar or (preferably) the same structure as itself.

Property • A *property* refers to the manner in which an entity (or a process) affects other entities (or other processes) in a certain situation (e.g., mass, position, length, name, velocity, etc.).

Q

Quantum Weirdness • The concept of *quantum weirdness* refers to the unintuitive results we see when looking at effects in the quantum world. Intuitively, we expect everything to act the way it does in our immediate, slow-moving and high-energy "macroworld." But for particles, this intuitive approach does not work, hence we call the quantum world "weird" in that regard.

Qubit • A *qubit* can be compared to a wave (as opposed to a discrete unit as with digital bits). Combining several qubits with a quantum gate results in an instant calculation of an interference pattern between the waves, which can be used to significantly speed up processing.

R

The Red Queen Hypothesis • "Red Queen" is a reference the novel *Through the Looking-Glass* where the protagonist needs to run just to keep in the same place. The *Red Queen Hypothesis* refers to a stable balance between two competing species that are interlocked in their evolution and where both are evolving, yet no side gets the upper hand. They are bound to each other in a way that the failure of one species is the success of the other. A gazelle failing to outrun a cheetah ultimately helps the cheetah to procreate.

Likewise, a cheetah failing to catch a gazelle helps the gazelle to procreate. Another example would be plants and caterpillars, with the plants producing more and more toxic leaves, while the caterpillars become more and more resistant to the toxin. Over time, both species become increasingly specialized in catching or evading each other.

Reversal of the Burden of Proof • Using the argument of the *reversal of the bur-* *den of proof,* you try to evade the necessity to give proof for your own arguments and instead present the opposite of your argument, and ask the other person for proof. The basic (and wrong) premise of the reversal of the burden of proof argument is that anything that cannot be disproven must be true. This is a fallacy because you often cannot prove a negative without being omniscient.

S

Science • *Science* is the formalized process of gaining new knowledge from observation, deducting new knowledge from existing knowledge, and checking existing knowledge for contradictions.

Selection (Evolution) • *Selection* is a process where some (or all) of a set of systems of similar structure are retained while the rest are discarded or destroyed. Which ones are retained and which ones are discarded depends on the relationship between the structure of the individual systems and the fitness landscape. In the example of the riddle screen, the screen lets small sand fall through while it "selects" larger stones.

Structure • A *structure* is a description of required properties, dependencies, and arrangement of a number of entities (e.g., cube-shaped).

System • A *system* is an aggregate with a definite structure (e.g., an ice cube is a system of frozen water molecules).

T

Theory of Evolution • The *theory of evolution* states that the process of evolution tends to create systems in each new generation that are better adapted to the environment (same or higher rate of procreation compared to the parent generation).

Time • *Time* is a measurement tool to put the speed of processes in relation to each other.

Quotation Sources

v: Krauss, *A Universe From Nothing*, p. 138f
xvii: Beagle, *The Last Unicorn*, p. 169
1: Arendt, *The Origins of Totalitarianism*, p. 474
13: Feynman, *Messenger Lectures, The Character of Physical Law*
29: Feynman and Leighton, *What Do You Care What Other People Think? Further Adventures of a Curious Character*, p. 237
34: Feynman, *Fun to Imagine*
40: Krauss, *A Universe From Nothing*, p. 151
53: Peikoff, *Understanding Objectivism*, p. 170
53: Bohm and Hiley, *The Undivided Universe*, p. 4-5
54: Bohm and Hiley, *The Undivided Universe*, p. 2
56: Bell, *Speakable and Unspeakable in Quantum Mechanics*, p. 170
57: Bohm, *Quantum Implications: Essays in Honour of David Bohm*, p. 93
58: Bell, *Speakable and Unspeakable in Quantum Mechanics*, p. 171
69: Feynman, *The Character of Physical Law*, p. 28
78: Bohm, *David Bohm and the Big Bang*
81: Krauss, *A Universe From Nothing*, p. 169
89: fortunesfool, *Aquinas and Evolution*, cf.
99: Sagan, Hanzlik, and Fahn, *Cosmos: Blues for a Red Planet*
100: Konner, *The Tangled Wing*, p. xviii
100: Mayr, *The Growth of Biological Thought*, p. 47
117: Darwin, *Letter To J. D. Hooker*
126: Dawkins, *The Blind Watchmaker*, p. 158
139: Peikoff, *Understanding Objectivism*, p. 170
153: Carroll, *Through the Looking-Glass*, p. 27
166: Calvin, *How Brains Think: Evolving Intelligence, Then And Now*, p. 1
168: Rand, *The Romantic Manifesto*, p. 36
185: Beagle, *The Last Unicorn*, p. 178
191: Mountain Dreamer, *The Invitation*, p. 1
195: Catmull and Wallace, *Creativity, Inc.: Overcoming the Unseen Forces That Stand in the Way of True Inspiration*, p. 114

Bibliography

Aboelsoud, N.H. *Herbal medicine in ancient Egypt.* In: *Journal of Medicinal Plants Research* Vol. *4(2)* (2010), pp. 82–86.

Abraham, Carolyn. *Sleep less, live longer, scientists say in new study.* [online; last accessed Dec 18th, 2017]. 2017. URL: https://www.theglobeandmail.com/life/sleep-less-live-longer-scientists-say-in-new-study/article22395327/.

AgileMethodology. *The Agile Movement.* [online; last accessed Apr 12, 2017]. 2008. URL: http://www.agilemethodology.org.

Ali, Ahmed Farag and Saurya Das. *Cosmology from quantum potential.* In: *Physics Letters B* 741.Supplement C (2015), pp. 276 –279. ISSN: 0370-2693. DOI: https://doi.org/10.1016/j.physletb.2014.12.057. URL: http://www.sciencedirect.com/science/article/pii/S0370269314009381.

Andersen, Anders et al. *Double-slit experiment with single wave-driven particles and its relation to quantum mechanics.* In: *Phys. Rev. E* 92 (1 2015), p. 013006. DOI: 10.1103/PhysRevE.92.013006. URL: https://link.aps.org/doi/10.1103/PhysRevE.92.013006.

Arendt, Hannah. *The Origins of Totalitarianism.* Mariner Books, 1973. ISBN: 978-0156701532.

Beagle, Peter S. *The Last Unicorn.* Roc Trade, 1991. ISBN: 045145052-3. URL: https://amzn.to/2pYj9oW.

— *The Last Unicorn.* Roc Trade, 1991. ISBN: 978-0451450524.

Beck, Kent. *Principles Behind the Agile Manifesto.* [online; last accessed Apr 12, 2017]. 2001. URL: http://agilemanifesto.org/principles.html.

Bell, J.S. *Speakable and Unspeakable in Quantum Mechanics.* Cambridge University Press, 1988. ISBN: 978-05-2136-869-8. URL: https://amzn.to/2EcepkN.

Bohm, David. *Bohm—Wholeness and the Implicate Order.* Routledge Classics, 1980. ISBN: 978-04-1528-979-5. URL: https://amzn.to/2uE2k8C.

— *David Bohm and the Big Bang.* [online; last accessed December 1st, 2017]. URL: http://blog.dbohm.com/tag/big-bang/.

— *Quantum Implications: Essays in Honour of David Bohm.* Routledge, 1991. ISBN: 978-0415069601. URL: https://amzn.to/2ImcjB8.

Bohm, David and Basil J. Hiley. *The Undivided Universe.* Routledge, 1995. ISBN: 978-0415121859.

Bradbury, Phillip. *Life in Life.* [online; last accessed Dec 18th, 2017]. 2012. URL: https://www.youtube.com/watch?v=xP5-iIeKXE8.

California, University of. *Open-source software for volunteer computing and grid computing.* [online; last accessed August 19th, 2012]. 2012. URL: http://boinc.berkeley.edu.

Calvin, William H. *A Brief History of the Mind: From Apes to Intellect and Beyond.* Oxford University Press, 2007. ISBN: 978-01-9518-248-4. URL: https://amzn.to/2GueemQ.

— *How Brains Think: Evolving Intelligence, Then And Now.* Basic Books, 1997. ISBN: 978-0465072781.

— *The Cerebral Code: Thinking a Thought in the Mosaics of the Mind.* A Bradford Book, 1996. ISBN: 978-0262531542. URL: https://amzn.to/2GsrGMc.

Campbell, Neil A. et al. *Biology.* 11th ed. Pearson, 2016. ISBN: 978-0134093413.

Carroll, Lewis. *Through the Looking-Glass.* Macmillan Children's Books, 2015. ISBN: 978-1447280002.

Catmull, Ed and Amy Wallace. *Creativity, Inc.: Overcoming the Unseen Forces That Stand in the Way of True Inspiration.* Random House, 2014. ISBN: 978-0812993011.

Cialdini, Robert B. *Influence—The Psychology of Persuasion.* HarperCollins, 2007. ISBN: 978-00-6124-189-5.

Colloca, Luana and Fabrizio Benedetti. *Placebos and painkillers: is mind as real as matter?* In: *Nature Reviews Neuroscience* 6 (2005), pp. 545 –552.

Darwin, Charles. *Letter To J. D. Hooker.* [online; last accessed December 1st, 2017]. 2012. URL: https://www.darwinproject.ac.uk/letter/DCP-LETT-4065.xml.

Dawkins, Richard. *Climbing Mount Improbable.* W. W. Norton and Company, 1997. ISBN: 978-0393316827. URL: https://amzn.to/2GswpcJ.

— *The Blind Watchmaker.* W. W. Norton & Company, 2015. ISBN: 978-0393351491. URL: https://amzn.to/2H4vNe8.

— *The Selfish Gene.* Oxford University Press, 1990. ISBN: 978-01-9286-092-7. URL: https://amzn.to/2EbI0eb.

Fernando, Chrisantha, K. K. Karishma, and Eörs Szathmáry. *Copying and Evolution of Neuronal Topology.* In: *PLOS ONE* 3.11 (Nov. 2008), pp. 1–21. DOI: 10.1371/journal.pone.0003775. URL: https://doi.org/10.1371/journal.pone.0003775.

Feynman, Richard. *Fun to Imagine.* [online; last accessed Dec 18th, 2017]. 1983. URL: http://www.bbc.co.uk/archive/feynman/.

— *Messenger Lectures, The Character of Physical Law.* [online; last accessed Dec 18th, 2017]. 1964. URL: http://www.cornell.edu/video/richard-feynman-messenger-lecture-7-seeking-new-laws.

Feynman, Richard P. *The Character of Physical Law.* Modern Library, 1994. ISBN: 978-0679601272. URL: https://amzn.to/2Ilewgs.

Feynman, Richard P. and Ralph Leighton. *What Do You Care What Other People Think? Further Adventures of a Curious Character.* W W Norton, 2008. ISBN: 978-03-9332-092-3. URL: https://amzn.to/2IlxT96.

Feynman, Richard P. and Jeffrey Robbins. *The Pleasure of Finding Things Out.* Basic Books, 2005. ISBN: 978-04-6502-395-0. URL: https://amzn.to/2GqWQPQ.

fortunesfool. *Aquinas and Evolution.* [online; last accessed August 3rd, 2012]. 2007. URL: http://serendip.brynmawr.edu/exchange/node/319.

Hanschen, Erik R. et al. *The Gonium pectorale genome demonstrates co-option of cell cycle regulation during the evolution of multicellularity.* In: *Nature Communications* 7 (Apr. 2016), pp. 11370+. ISSN: 2041-1723. DOI: 10.1038/ncomms11370. URL: http://dx.doi.org/10.1038/ncomms11370.

He, Dongshan, Dongfeng Gao, and Qing-yu Cai. *Spontaneous creation of the universe from nothing.* In: 89 (Apr. 2014).

Ingraham, Paul. *Why does Pain Hurt?* [online; last accessed Nov 12, 2017]. 2010. URL: https://www.painscience.com/articles/why-does-pain-hurt-so-much.php.

Konner, Melvin. *The Tangled Wing.* Holt Paperbacks, 2003. ISBN: 978-0805072792.

Kozloski, James. *Closed-Loop Brain Model of Neocortical Information-Based Exchange.* In: *Frontiers in neuroanatomy* 10 (2016), p. 3. ISSN: 1662-5129. DOI: 10.3389/fnana.2016.00003. URL: http://europepmc.org/articles/PMC4716663.

Krauss, Lawrence M. *A Universe From Nothing.* Free Press, 2012. ISBN: 978-14-5162-445-8. URL: https://amzn.to/2In0GKp.

Lode, Clemens. *Philosophy for Heroes: Act.* Clemens Lode Verlag e.K., 2018. ISBN: 978-39-4558-623-5.

— *Philosophy for Heroes: Continuum.* Clemens Lode Verlag e.K., 2017. ISBN: 978-39-4558-622-8.

— *Philosophy for Heroes: Epos.* Clemens Lode Verlag e.K., 2019. ISBN: 978-39-4558-624-2.

— *Philosophy for Heroes: Knowledge.* Clemens Lode Verlag e.K., 2016. ISBN: 978-39-4558-621-1. URL: https://amzn.to/2H4wx2U.

— *Scrum Your Jira! Your Waterfall Organization Transformed Into Agile Multidisciplinary Teams.* Clemens Lode Verlag e.K., 2017. ISBN: 978-3945586655. URL: https://amzn.to/2uE3NMa.

Loizeau, Nicolas. *Game of life: Programmable Computer*. [online; last accessed Dec 18th, 2017]. 2016. URL: https://www.youtube.com/watch?v=8unMqSp0bFY.

Mayr, Ernst. *The Growth of Biological Thought*. Harvard Univ Press, 1985. ISBN: 978-0674364462.

Mithen, Steven. *The Singing Neanderthals—the Origins of Music, Language, Mind, and Body*. Harvard University Press, 2007. ISBN: 06-7402-559-8. URL: http://amzn.to/2jMGusK.

Mountain Dreamer, Oriah. *The Invitation*. HarperOne, 2006. ISBN: 978-0061116711.

Neumann, John von. *Theory of Self-reproducing Automata*. University of Illinois Press, 1967. ISBN: 978-0252727337.

Ohno, Taiichi. *Ask why five times about every matter*. [online; last accessed Apr 12, 2017]. 2006. URL: http://www.toyota-global.com/company/toyota_traditions/quality/mar_apr_2006.html.

Peikoff, Leonard. *Objectivism: The Philosophy of Ayn Rand*. Dutton, New York U.S.A., 1991. ISBN: 05-2593-380-8. URL: https://amzn.to/2q1mns6.

— *Understanding Objectivism*. NAL Trade, 2012. ISBN: 978-04-5123-629-6. URL: http://amzn.to/2jMBLYf.

PMI. *A Guide to the Project Management Body of Knowledge - Fifth Edition*. Project Management Institute, 2013. ISBN: 978-1935589679. URL: http://amzn.to/2owDPTe.

Rand, Ayn. *Address to the Graduating Class of the United States Military Academy at West Point New York*. [online; last accessed Apr 12, 2017]. 1974. URL: http://fare.tunes.org/liberty/library/pwni.html.

— *The Romantic Manifesto*. Signet, 1971. ISBN: 978-04-5114-916-9.

Rand, Ayn, Harry Binswanger, and Leonard Peikoff. *Introduction to Objectivist Epistemology*. Expanded 2nd ed. New American Library, New York, N.Y., 1990. ISBN: 04-5201-030-6. URL: https://amzn.to/2IkrMlf.

Ray, Tom. *Tierra Photoessay*. [online; last accessed Nov 12, 2017]. URL: http://life.ou.edu/pubs/images/.

— *Welcome to the Tierra home page*. [online; last accessed August 19th, 2012]. 2008. URL: http://life.ou.edu/tierra/.

Sagan, Carl. *The Demon-Haunted World: Science as a Candle in the Dark*. Ballantine Books, 1997. ISBN: 978-03-4540-946-1. URL: https://amzn.to/2IIyQOI.

Sagan, Carl, Jaromir Hanzlik, and Jonathan Fahn. *Cosmos: Blues for a Red Planet*. USA, 2002.

Scientist, New. *First replicating creature spawned in life simulator*. [online; last accessed Dec 18th, 2017]. 2010. URL: https://www.newscientist.com/article/mg20627653-800-first-replicating-creature-spawned-in-life-simulator/.

Shay, Jerry W and Woodring E Wright. *Role of telomeres and telomerase in cancer*. In: *Seminars in Cancer Biology 21.6* (2011), 349 −353. URL: http://www.ncbi.nlm.nih.gov/pmc/articles/PMC3370415/.

Stackexchange. *Build a digital clock in Conway's Game of Life*. [online; last accessed Dec 18th, 2017]. 2017. URL: https://codegolf.stackexchange.com/questions/88783/build-a-digital-clock-in-conways-game-of-life.

TechTarget. *What is Scrum?* [online; last accessed Apr 12, 2017]. 2007. URL: http://searchsoftwarequality.techtarget.com/definition/Scrum.

Index

Is what you are looking for not here? Send us a quick message and help us to improve the index: index2@lode.de

An Important Final Note

Writers are not performance artists. While there are book signings and public readings, most writers (and readers) follow their passion alone in their homes.

*What applause is for the musician, **reviews** are for the writer.*

Books create a community among readers; you can share your thoughts among all those who will or have read the book.

Leave a thoughtful honest review and help me to create such a community on the platform on which you have acquired this book. *What did you like, what can be improved? To whom would you recommend it?*

Thank you, also in the name of all the other readers who will be able to better decide whether this book is right for them or not! A positive review will increase the reach of the book, a negative review will improve the quality of the next book. I welcome both!

To every man is given the key to the gates of heaven; the same key opens the gates of hell.

—Buddhist proverb

Made in the USA
Middletown, DE
26 April 2019